Predicting Semiconductor BusinessTrends After Moore's Law

Dr. Walden Rhines

A SemiWiki.com Project

Predicting Semiconductor Business Trends After Moore's Law

Author: Walden Rhines

ISBN: 9798609900760

BISAC: Business & Economics / General

Acknowledgments

I'd like to offer my sincere thanks to the many people in Mentor Graphics (now Mentor, a Siemens Business) who made this book possible.

Over the last twenty years, Merlyn Brunken, Manager of Market Research for Siemens PLM has performed much of the analyses used to create the book, as well as participating in the formulation of conclusions. His natural sense for industry trends, knowledge of credible data sources and analytical skills uncovered trends and relationships in semiconductor and EDA markets that have been overlooked by many qualified analysts. His impeccable honesty provided the comfort that data would be accurately reported and that conclusions would be based upon reliable sources of data.

Laura Parker, Senior Manager, Technology Leadership, prepared a large share of the slides and participated in many of the strategic discussions as we formulated the themes and analyzed the data that make up the book. Kathy Brost, Technical Librarian for Mentor, a Siemens Business, was able to find data that provided insight in dozens of presentations. She compiled this data into many of the slides in the book. Laura Olson performed research for many of the issues discussed in the book. Anne Cirkel, Senior Director of Marketing, provided insights in the preparation of some of the presentations and made many of the slides comprehensible in different contexts.

Finally, Lisa Blanchard, Executive Assistant, managed the administration for hundreds of meetings, presentations and calls related to the collection and analysis of data and the preparation of presentations.

Table of Contents

Introduction

In 1952, AT&T sold licenses to patents and basic know-how for their newly developed solid-state transistor technology to any buyer willing to pay $50,000. As a result, the companies who chose to commercialize this technology competed on a level playing field with no initial competitive barriers such as patents or existing market share. They created what soon became the most significant example of a free market business operating in a world economy. Regulations for this new industry didn't exist and the new companies created a hotbed of new ideas, new business approaches and financial growth. It was the "Wild West" of business. As a result, the semiconductor industry today provides the most significant example in recent history of free economics in worldwide commerce.

Without a formal licensing process, IBM's development of the Winchester disk drive had a similar effect beginning in 1956. Over the next thirty years, the number of companies competing in the hard disk drive business peaked at eighty-five. Clayton Christensen of Harvard University did a study of the disk drive industry because it could be analyzed using nearly ideal conditions of supply, demand and free market economics (see *Christensen, Clayton, "The Innovator's Dilemma: When New Technologies Cause Great Firms to Fail", Harvard Business Review Press, May 1, 1997.)*

He used disk drive companies as a surrogate for other industries in the same way that biological researchers use fruit flies. Fruit flies are born, mature, reproduce and die in 24 hours so you can study biological effects over many generations. Christensen's thesis was that the disk drive industry provided a research vehicle similar to fruit flies in that these companies were founded, grew and went out of business in a very short period of time.

The semiconductor industry exhibited life cycles that were longer than the disk drive industry but had the same free market characteristics. Over time this unfettered competition followed trends in a worldwide market that could be quantified and used to predict the future. Over the past forty years or more, I've collected data and made presentations showing how the actual economics and technology of the semiconductor industry can be used to predict its future direction and magnitude. This book is built upon excerpts of presentations made during the last thirty years that analyze the business and technology of the semiconductor industry. In most cases, the figures in the book are copies of the original slides as they were presented during one or more of those presentations. In general, they show how predictable the semiconductor industry has been. They should also provide insight into the future of the industry.

—Dr. Walden Rhines, December 2019

Tools for Predicting Semiconductor Trends

Chapter 1: Understanding the Learning Curves

Learning curves provide predictable cost or revenue per transistor. Figure 1 is the most basic of all the predictable parameters of the semiconductor industry, even more so than Moore's Law. It is the learning curve for the transistor. Since 1954, the revenue per transistor (and presumably the cost per transistor, if we had the data from the manufacturers) has followed a highly predictable learning curve. Before Moore's Law, the learning curve provided a guiding light for the semiconductor industry. Texas Instruments used it for strategic advantage and shared its data with Boston Consulting Group who published a book called "Perspectives on Experience"[1].

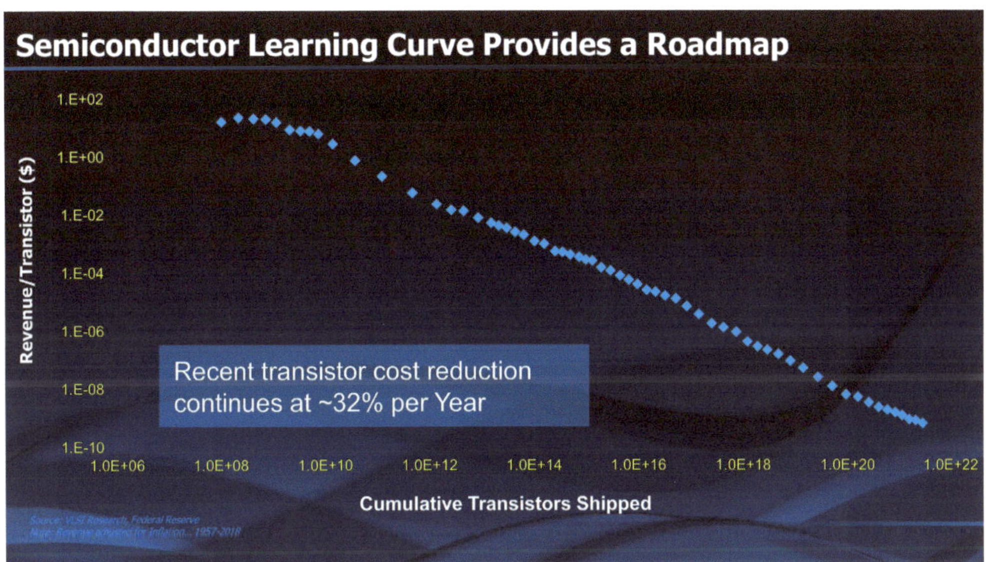

Figure 1. Learning curve for the transistor from 1954 to 2019.

In the days of germanium and silicon discrete transistors, companies like TI could use the learning curve, for example, to predict what the unit cost would be after 100,000 units were produced, based upon their own known actual cost per unit of the first 1,000 units produced. They could then price the particular transistor product at a loss initially to gain leading market share and therefore achieve higher profitability and market influence when they reached future high unit volume sales.

TI didn't create the technology of learning curves. It was developed in 1885[2] and has been used in industries like aviation, even before the transistor was invented, to predict the future cost per airplane when a certain cumulative unit volume was achieved. TI's unique approach for semiconductors lay in the use of the learning curve to drive a pricing strategy early in the life of a new component.

Figure 2 shows how the learning curve works. The vertical axis is the logarithm of the cost per unit of anything that is produced. The product can be a good or service; anything that benefits from the experience of doing the same thing, or making the same product, again and again. Published learning curves typically use the revenue per unit because companies are unwilling to divulge their cost data. The companies, however, know their costs and, over the history of the semiconductor industry, have used that data to strategically position themselves against competition.

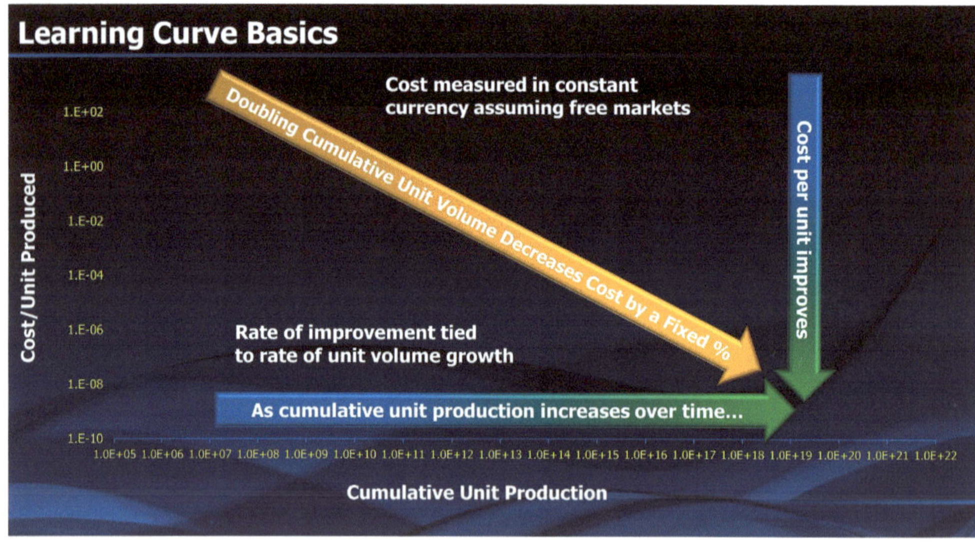

Figure 2. The learning curve is a log/log plot of cost per unit vs cumulative units manufactured.

The horizontal axis of the learning curve is the logarithm of the cumulative number of units of a product or service that have been produced throughout history. When the data is plotted, it results in a straight line with a downward slope. Cost per unit decreases monotonically as we develop more experience, or "learning".

The IC revenue growth rate shown by the wide yellow line is the average percent through 2015 plus projections made by analysts at the time of this presentation in 2016. As the cumulative number of units produced becomes very large, the time required to double the cumulative volume becomes longer, so the time required to reduce cost by a fixed percentage also increases. Every time the cumulative number of units produced doubles, the learning curve plot reflects a decrease in the cost per unit by a fixed percentage. The percentage is different for different products but tends to be similar across a broad range of products in an industry like semiconductors.

Many efficiencies contribute to reduction in cost (and therefore revenue) of a product as the cumulative number of units produced increases. For example, depreciation and development costs can be amortized over a larger unit volume of the product.

Learning curves can be applied to any good or service where the cost per unit of production can be measured. We are just not as aware of the phenomenon today because the measurement applies only when cost is measured in constant currency. A deflator must therefore be applied to the cost numbers to account for the portion of inflation that is caused by governmentally driven inflation.

In addition, the learning curve only applies in free markets. Tariffs, trade barriers, taxes and other costs must be removed before actual cost comparisons can be made. The reason that learning curves have been so valuable in the semiconductor industry is that it is one of the few industries that has operated for over sixty years in a relatively free worldwide market, with minimal regulation and tariffs as well as a very low cost of freight between regions.

One of the great things about semiconductor learning curves is that they will be applicable as long as transistors, or equivalent switches, are produced. While Moore's Law is quickly becoming obsolete, the learning curve will never be. What will happen, however, is that the cumulative number of transistors produced will stop moving so quickly to the right on the logarithmic scale. Then the prices will not decrease as rapidly as they have in the past. The visible effect of improved learning will diminish.

At some point, monetary inflation will be larger than the manufacturing cost reduction and transistor unit prices may actually increase with time in absolute dollars even though they are decreasing in constant currency. In the meantime, the learning curve is a useful guidepost for predicting the future. Currently, in 2019, the revenue per transistor is decreasing about 32% per year.

Those who purchase microprocessor or "system on chip" (SoC) components may recognize that, in 2017, the price per transistor is decreasing at a slower rate than 32% per year. Figure 3 explains this. The 32% number applies to the total of all semiconductor components produced in 2017. However the cost per transistor is made up of different kinds of semiconductor components — memory, logic, analog, etc.

It becomes apparent from Figure 3 that the semiconductor industry is producing far more transistors in discrete memory components, particularly NAND FLASH nonvolatile memories, than in other types of semiconductors. When the memory learning curve (consisting mostly of NAND FLASH and DRAM) is separated from the non-memory learning curve, it is evident that cost per transistor and cumulative unit volume for memory are way ahead of non-memory.

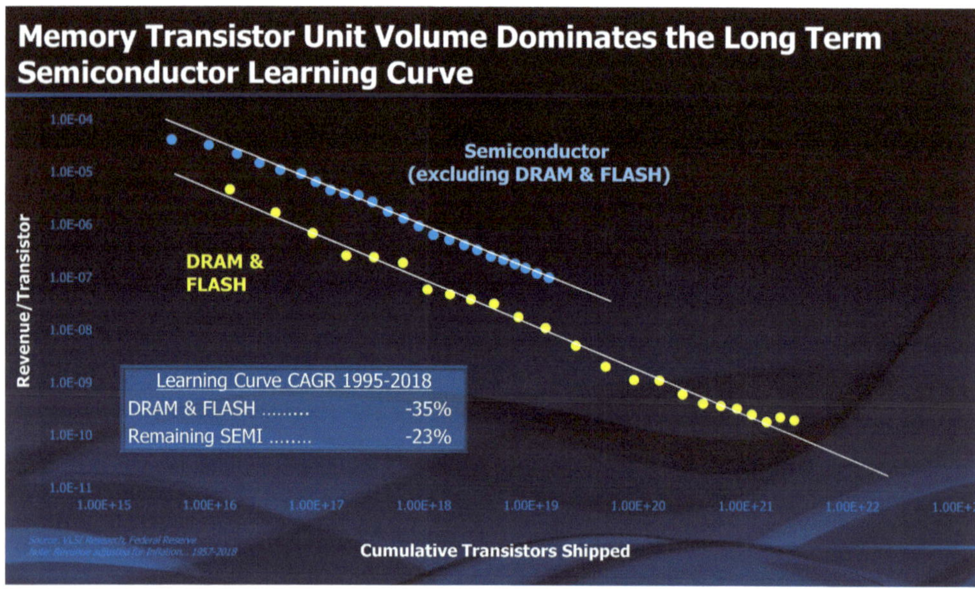

Figure 3. Cumulative unit volume of transistors used in memory components is increasing much faster than unit volume of transistors in other types of chips.

That's okay because the learning curve doesn't specify how the decreasing cost per transistor is achieved — only that it will happen as a function of cumulative transistors produced.

Another aspect of interest in Figure 3 is the set of data points near the end of the curve that were generated by data from 2017 and 2018. The data points are above the learning curve trend line. How can this happen if the learning curve is a true law of nature? Very simply, the period from 2016 through 2018 was one of memory shortages, particularly DRAM. Prices per transistor increased instead of decreasing because market demand exceeded supply.

Won't this cause a long- term deviation from the learning curve? No. Whenever a market supply/demand imbalance occurs, the cost per transistor moves above or below the long-term trend line of the learning curve. This is always a temporary move. When supply and demand come back in balance, the cost per transistor will move to the other side of the learning curve.

Area generated above the learning curve will normally be compensated by a nearly equal area below the learning curve and vice versa. This is another useful benefit of the learning curve because it allows us to predict the general trend of future prices even when short term market forces cause a perturbation.

While I've focused on transistors in this discussion of learning curves, it should be noted that we could just as easily use electrical "switches" as our unit of measure. The same learning curve would then work for mechanical switches, vacuum tubes and transistors as seen in Figure 5 of Chapter 3.

This figure also shows another attribute of the learning curve. In this case, the metric on the vertical axis is revenue per MIP (or millions of computer instructions per second) for various types of electrical switches. Learning curves can be used to predict improvements in performance, reliability (in FITS), power dissipation and many other parameters that benefit from the cumulative unit volume of production experience.

Learning curves also provide a useful tool for predicting "tipping points" for new technology adoption. A good example is the introduction of "compression technology" in the semiconductor test industry in 2001. In hindsight, a major innovation like this was inevitable just by examining the learning curve for the cost of testing a transistor in an integrated circuit (Figure 4).

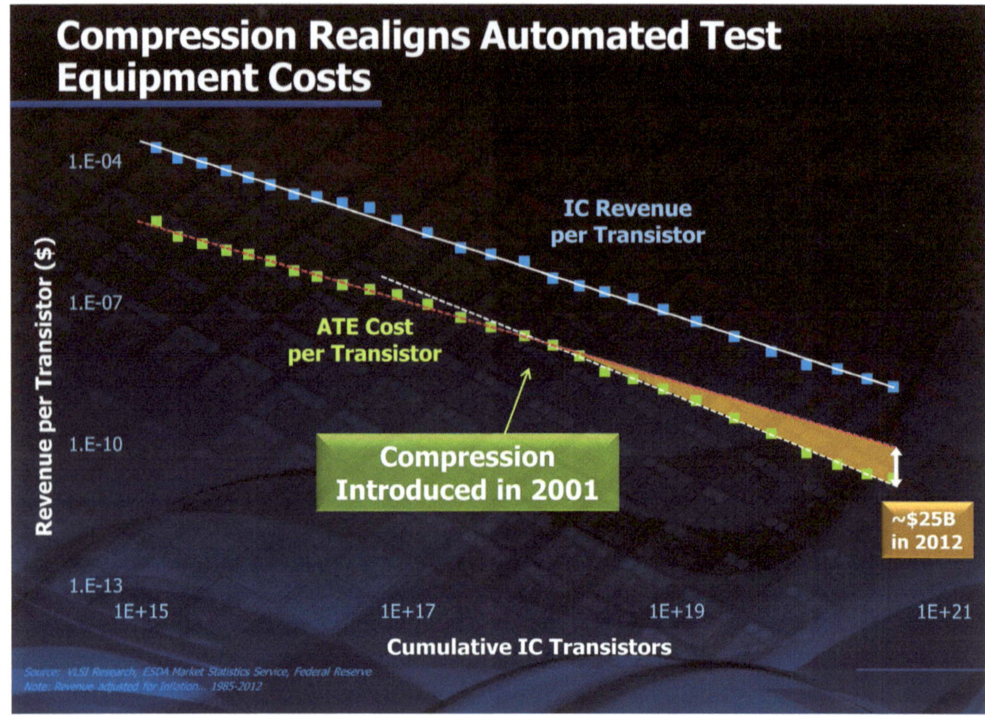

Figure 4. Until 2001, reduction in the revenue per transistor of the automated test equipment industry was decreasing at a slower rate than the transistors produced by their customers, the semiconductor component industry.

The revenue of the ATE industry directly impacts the cost of test for its customers. The ATE revenue learning curve, which is a "cost" learning curve for the semiconductor industry, was not parallel to the silicon transistor revenue learning curve and had a less steep slope.The ATE industry should have seen that change was inevitable. Pat Gelsinger, in his International Test Conference Keynote address in 1999 highlighted his prediction that "in the future, it may cost more to test a transistor than to manufacture it". Such a prediction would have occurred had it not been for compression technology (also called "embedded deterministic test") which started out in 2001 with a 10X improvement in the number of "test vectors" required to achieve the same level of test and then progressed to nearly 1000X by 2018[3].

Introduction of "embedded deterministic test", or test compression, in 2001 significantly reduced the number of testers required and, by 2012, reduced the revenue of the ATE industry by $25B per year.

1 Boston Consulting Group, *"Perspectives on Experience"*, 1970, Boston, MA

2 https://en. wikipedia. org/wiki/Learning_curve#In_machine_learning

3 Rajski, J. , Tyszer, J. , Kassab, M. and Mukherjee, N. , *"Embedded Deterministic Test"*, *IEEE Transactions on Computer-Aided Design of Integrated Circuits and Systems* (Volume: 23 , Issue: 5 , May 2004)

Chapter 2: Constants of the Semiconductor Industry

In the mid 1980s, Tommy George, then President of Motorola's Semiconductor Sector, pointed out to me that the semiconductor revenue per unit area had been nearly constant throughout the history of the industry including the period when germanium transistors made up a large share of semiconductor revenue. I began tracking the numbers at that time and continue to do so today. So far, it's still approximately true.

If you are making a decision about a capital investment in semiconductor manufacturing, or even an investment decision for the development of a new device, this is a remarkably useful parameter to test the wisdom of your investment. Figure 1 shows revenue per unit area data for the last twenty-five years (since I didn't keep my records before that time). There are many possible explanations for why this empirical observation should be approximately correct.

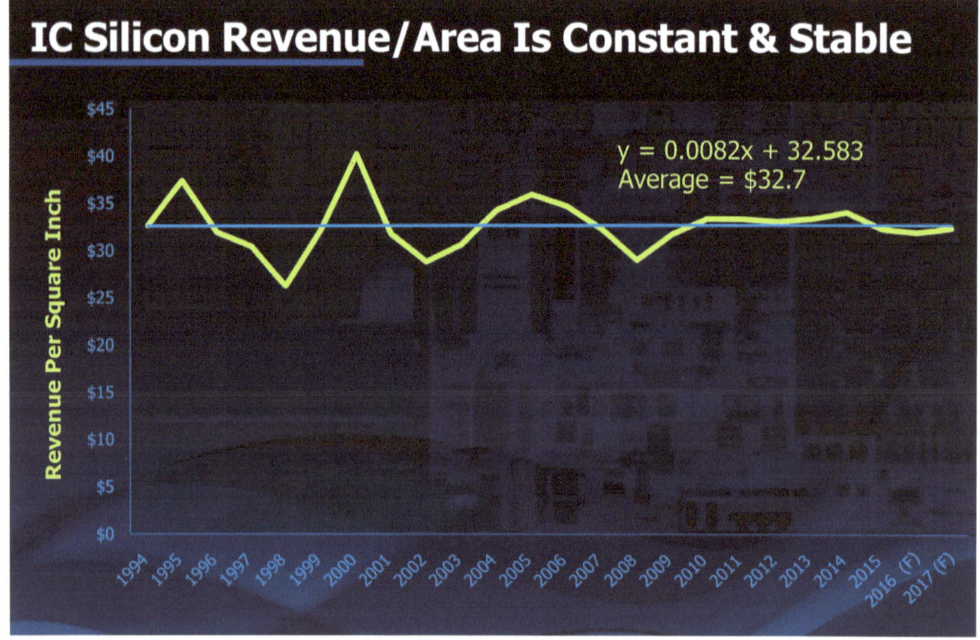

Figure 1. Revenue per unit area of silicon or germanium has been a long term constant of the semiconductor industry.

One of those explanations is the fact that semiconductor revenue and semiconductor manufacturing equipment costs, materials, chemicals and even EDA software costs all follow learning curves based upon the number of transistors cumulatively produced through history. Semiconductor revenue follows a learning curve that is parallel to the learning curves for all input costs to the design and production of semiconductors and is decreasing on a per transistor basis by more than 30% per year (Figures 2 through 6).

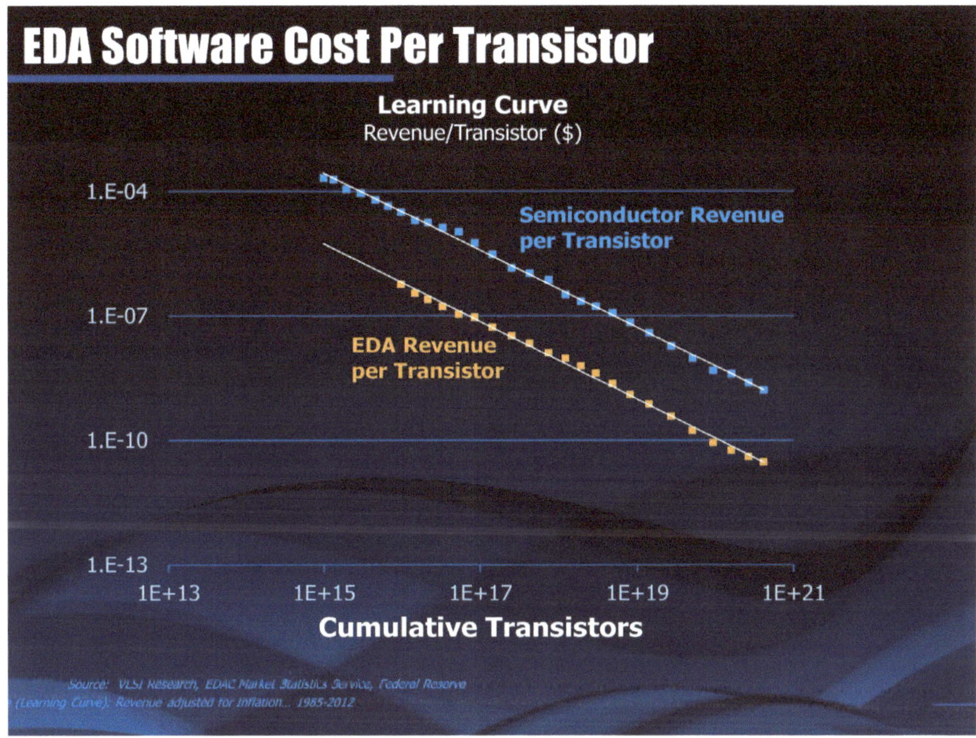

Figure 2. Learning curve for transistors and EDA software.

These input costs refer to all transistors produced in each incremental period. Manufacturing costs increase as we shrink feature sizes to the next node but the average cost per transistor is based upon manufacturing of integrated circuits that include very mature processes as well. The increasing manufacturing cost at each node is offset by both the increasing number of transistors per unit area and the longer term decreasing input costs of materials, equipment and software.

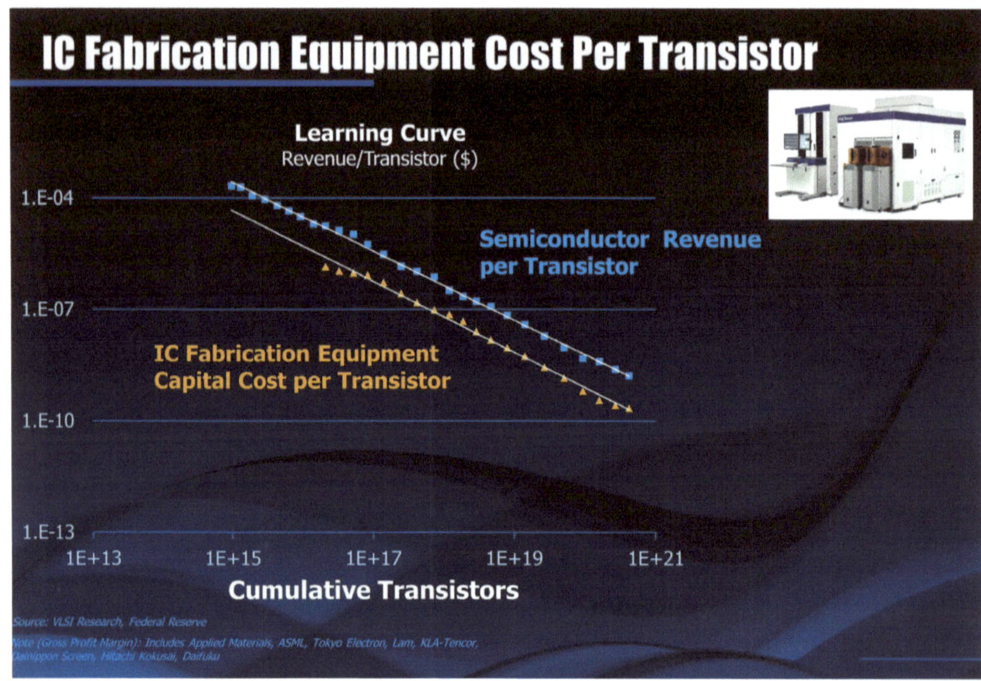

Figure 3. Learning curve for front end fabrication equipment.

The cost per transistor and the cost to process a fixed area of silicon therefore decrease at a nearly constant rate with the same slope which is also the same decreasing slope as the revenue per transistor. The ratio between revenue and area therefore stays approximately the same.

Figure 2 provides another observation that most customers of electronic design automation (EDA) software find surprising. I've found that most EDA customers think that the EDA industry charges too much for its software and doesn't feel the same pressure to reduce costs that is felt by its customers, the providers of chips. The learning curve for EDA software refutes this. The number of transistors sold by the semiconductor industry is a published number each year. So is the total revenue of the semiconductor industry and the EDA industry. When the EDA total available market (TAM) is divided by the number of transistors produced, we obtain the EDA software cost per transistor.

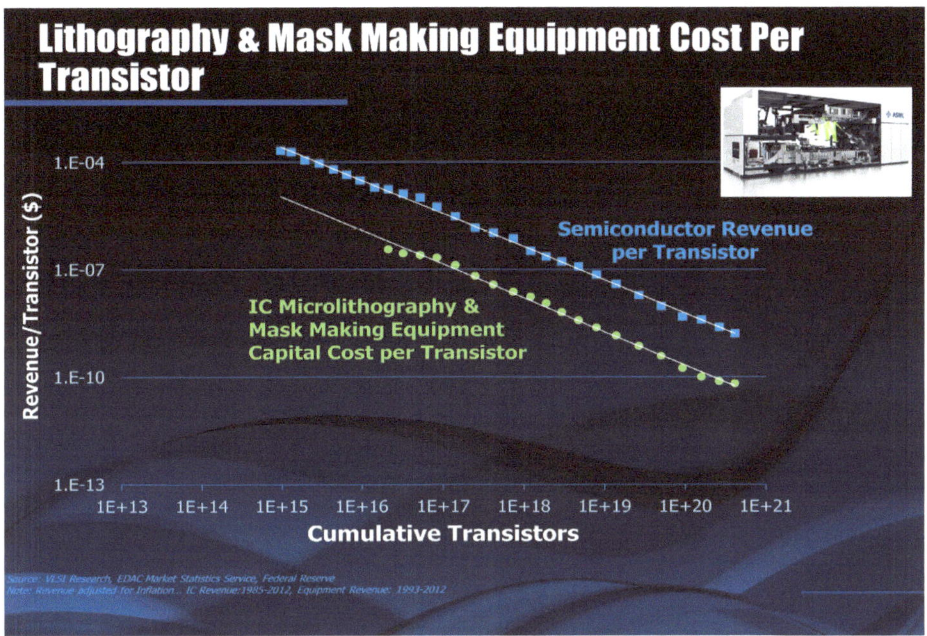

Figure 4. Learning curve for lithography and photomask making equipment.

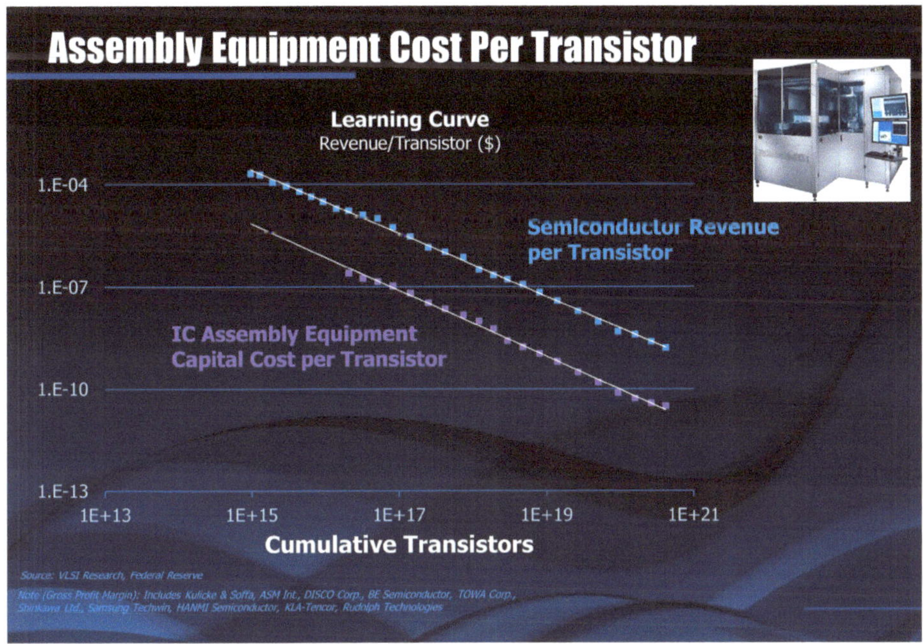

Figure 5. Learning curve for semiconductor assembly equipment.

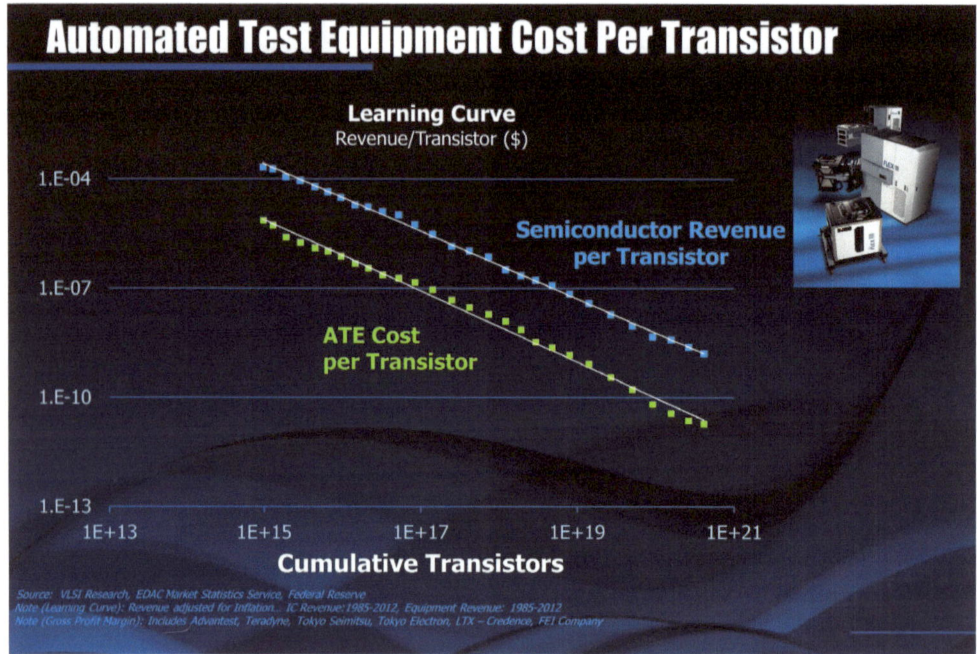

Figure 6. Learning curve for semiconductor automated test equipment.

This then shows that the EDA industry is reducing the cost of its products at the same rate as the semiconductor industry. That is as it must be. If the EDA industry doesn't keep its learning curve parallel to the semiconductor industry learning curve, then the cost of EDA software as a percent of semiconductor revenue would increase and there would have to be cost reductions elsewhere in the semiconductor supply chain to offset it.

As it is, EDA software costs are about 2% of worldwide semiconductor revenue (Figure 7) and this percentage has been relatively constant for the last twenty-five years. This is also a fixed percentage of worldwide semiconductor research and development (Figure 8) which has been a relatively constant 14% for more than thirty years.

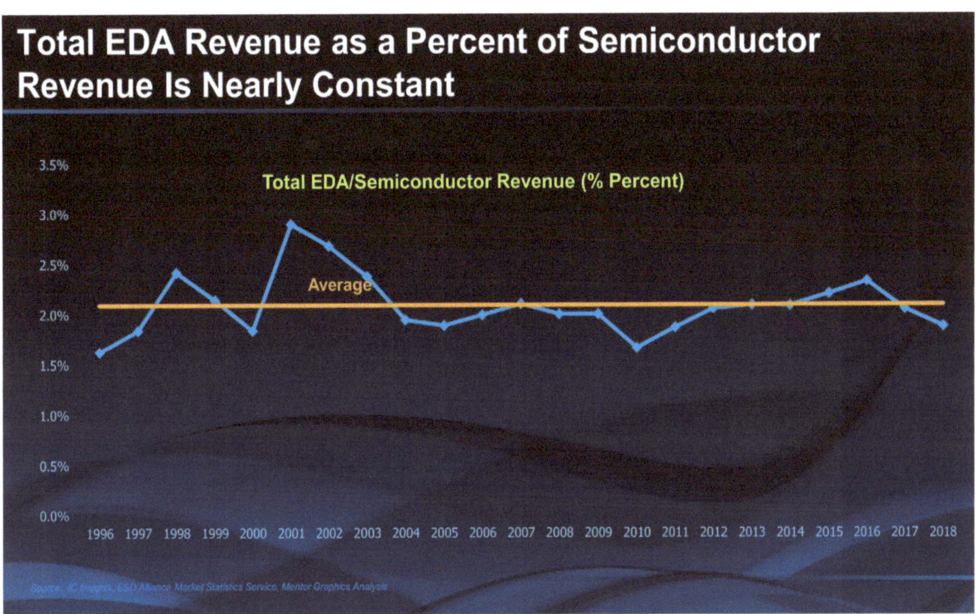

Figure 7. EDA revenue as a percent of semiconductor revenue.

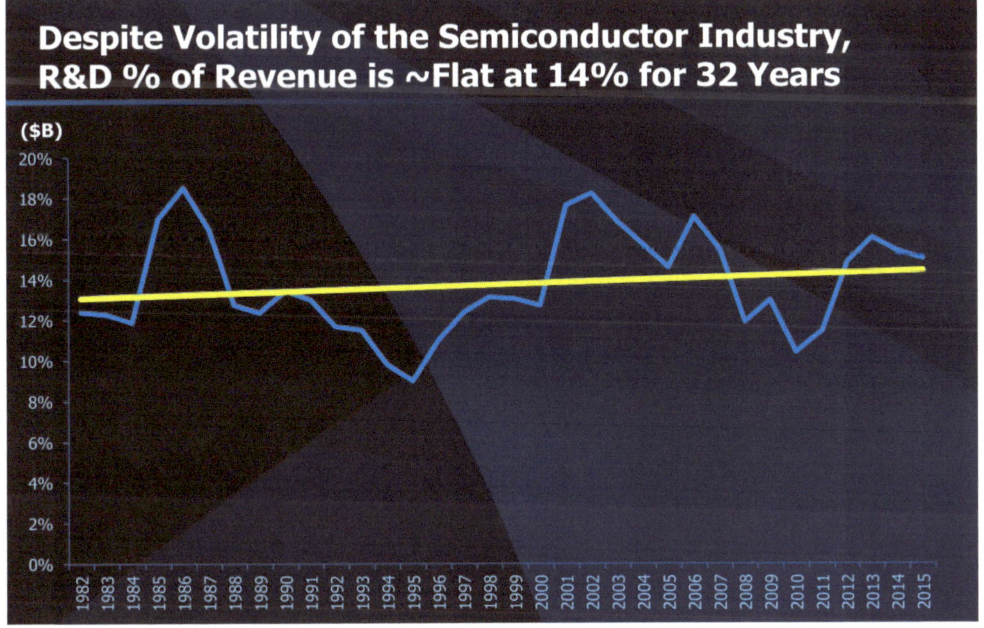

Figure 8. Semiconductor Research and Development as a percent of semiconductor industry revenue.

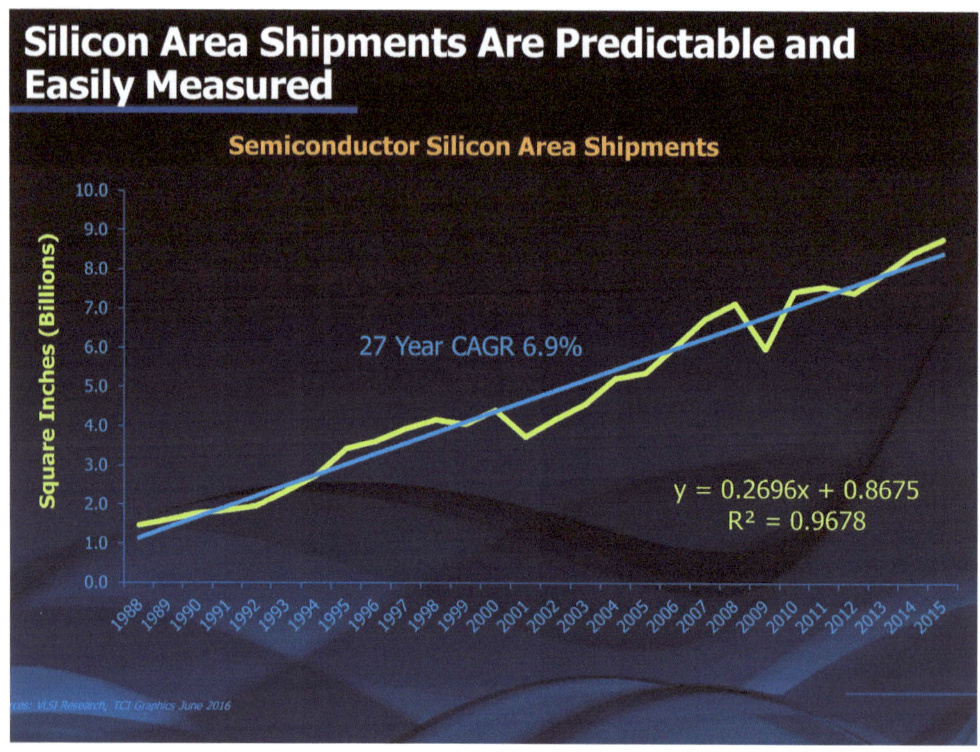

Figure 9. Area of silicon produced each year.

Figure 9 shows the annual production of silicon measured by area. This measurement follows a long term predictable curve. Actual data moves above and below the trend line as companies over-invest in capacity when demand is strong and under-invest in periods of market weakness. Investing counter-cyclically seems like a brilliant strategy but it's very difficult to execute because semiconductor recessions force companies to squeeze capital budgets and to under-invest just when they should be investing. Even so, this graph is useful because silicon area production is one thing that is predictable at least one year ahead.

We know approximately how much silicon area the existing wafer fabs are capable of producing and we are aware of the new wafer fabs that will be starting production in the coming year. Wafer fabs that are pulled out of service are a small percentage of the total, especially during strong market periods, so next year's silicon area is known fairly accurately. Since market demand is not known, shortages and periods of excess capacity lead to magnified price changes as the capacity grows monotonically.

But the growth or decrease of revenue in the coming year tends to be predictable when market supply and demand are reasonably balanced. We know the revenue per unit area of silicon. We also know the area of silicon that will be produced next year. Multiplying these two numbers gives us the revenue for next year. That's a useful number.

Figure 10 shows how the annual semiconductor revenue correlates with a calculation based upon silicon area multiplied by the predicted ratio of revenue per unit area of silicon. I find the correlation to be both remarkable and very useful. The IC revenue growth rate shown by the wide yellow line is the average percent through 2015 plus projections made by analysts at the time of this presentation in 2016.

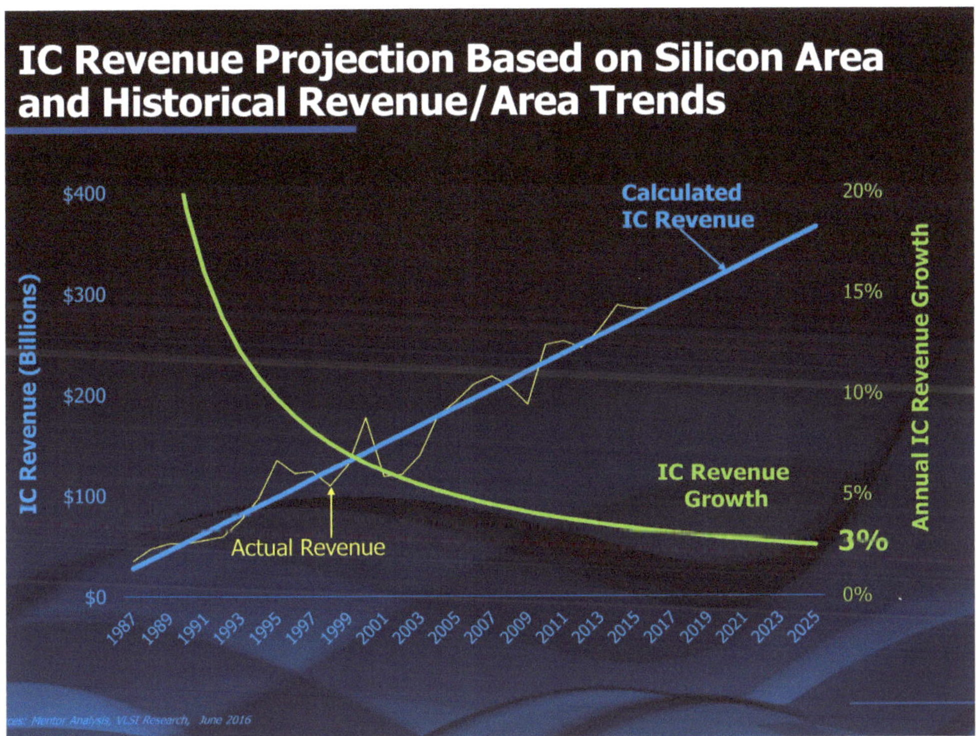

Figure 10. Integrated circuit revenue vs calculation from silicon area and the revenue per unit area ratio.

Semiconductor units shipped per year is also predictable (Figure 11). This data from VLSI Technology covers the period since 1994. While modest deviations do occur in years of severe recession or accelerated recovery, the long term trend is apparent and predictable. If you are purchasing capital for the long term, especially for assembly and test equipment, this data is particularly useful. Unit volume served as a cornerstone of semiconductor forecasts at the time I joined the industry in 1972.

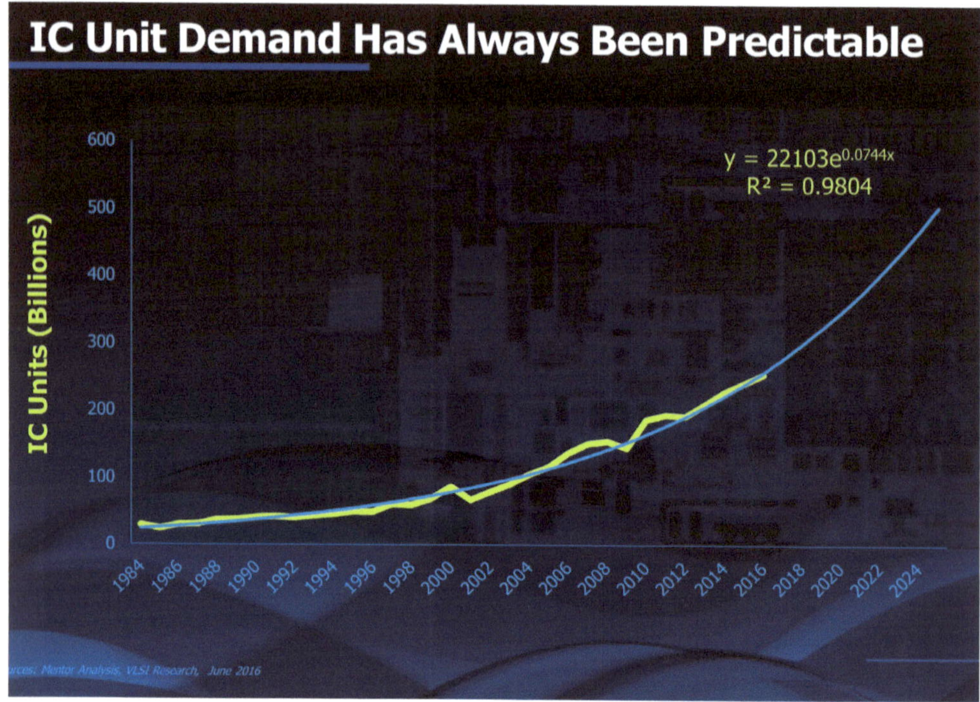

Figure 11. Integrated circuit annual unit volume of sales.

As far as I know, the unit volume had grown every MONTH since the start of the industry and it continued to do so until December of 1974 when the oil shock caused implosion of the market and semiconductor volume fell precipitously.

One of the most common errors in semiconductor forecasting occurs when forecasters look only at revenue, ignoring the variability of price in the long term trend. The unit volume is stable and predictable but the price is not. At the Symposium on VLSI Technology in Hawaii in 1990, Gordon Moore and Jack Kilby were present and we all commiserated about the death of Bob Noyce (who was Chairman of SEMATECH at the time) the day before the conference started.

Despite his grief, Gordon went ahead with his presentation the next day highlighting what might be referred to as Moore's Second Law (Figure 12), although it never caught on (for good reason). Gordon graphed the average selling price (ASP) of semiconductor components over their lifetimes, especially memory components. His conclusion was that semiconductor components that start out at higher prices will eventually cost $1.00.

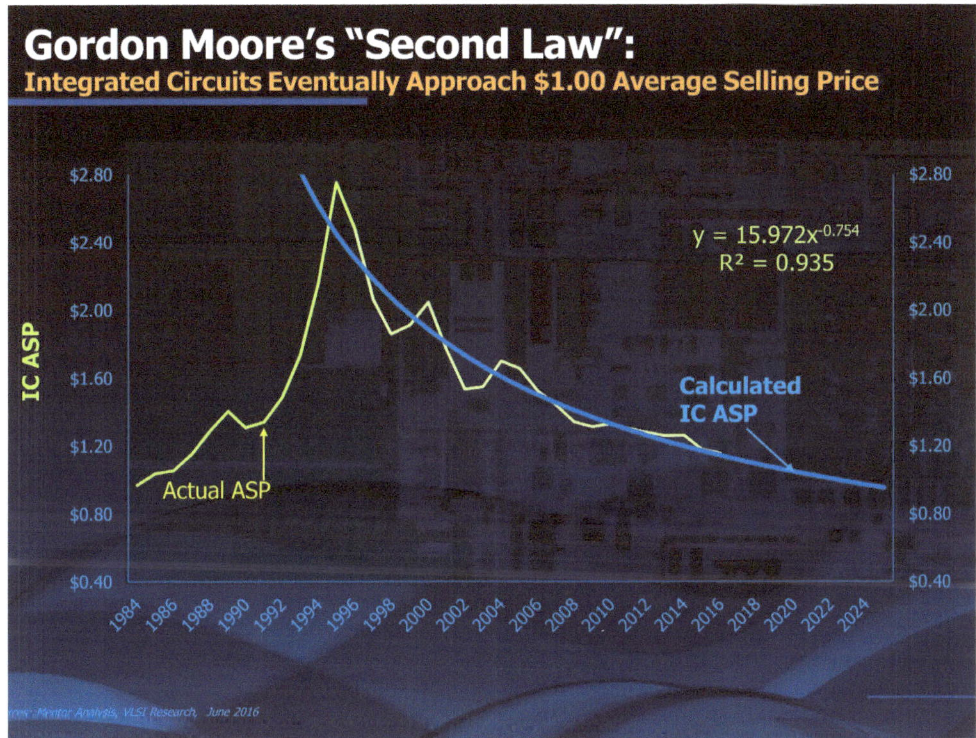

Figure 12. Average selling prices (ASPs) of semiconductor components.

Figure 13 shows the data since 1984. While the current trend and the distant history suggest that Gordon may have been right, this trend reveals major interruptions, the most notable of which was the DRAM shortage that occurred when Windows '95 was introduced in 1995. That drove up ASPs and we have been slowly trending down ever since then toward the $1.00 asymptote. The $1.00 price point should never be reached because there will always be newer components coming into the market but Gordon's hypothesis is certainly interesting if not compelling.

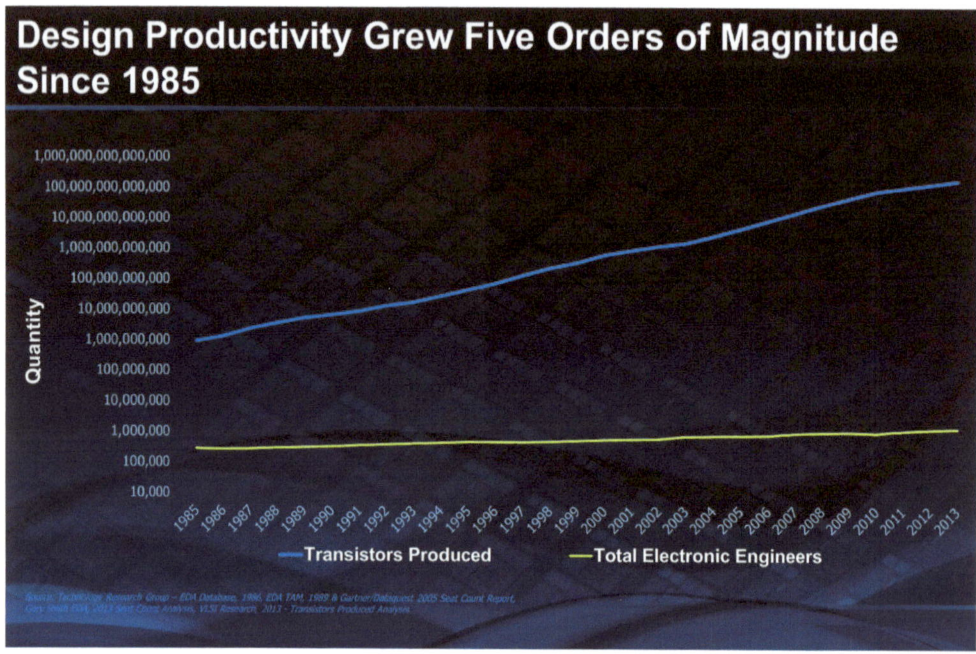

Figure 13. Transistors produced per electronic engineer.

One more interesting statistic is the number of transistors produced per engineer each year (Figure 13). This is a quasi-measure of design productivity that reflects both the growing number of transistors per chip as well as the increasing volume of chips that have been sold each year. By this measure, productivity has increased five orders of magnitude since 1985.

Chapter 3: Moore's Law is Unconstitutional

(Adapted from a presentation first given under this title in 1989 and subsequently expanded in presentations over a period of nearly thirty years).

In 1965, Gordon Moore, then R&D Manager for Fairchild Semiconductor, published a paper in "Electronics" magazine predicting the trend for semiconductors in the next ten years. He showed a graph of the number of components in the largest chips in each of the last four years that followed a straight line when plotted with a Y-axis that was the base two logarithm of the number of components (transistors, capacitors, resistors or diodes) and the horizontal axis was time. The number of components had doubled every year (Figure 1).

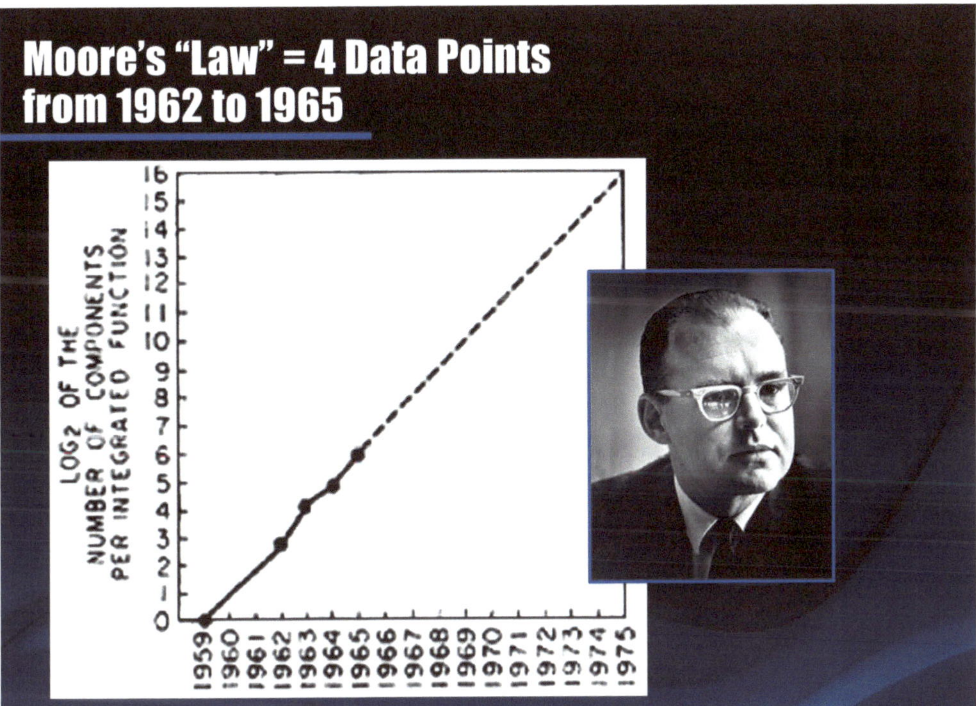

Figure 1. First presentation of Moore's Law in 1965.

This graph became known as "Moore's Law" and has been extrapolated for more than fifty years. It is not a "law". It is an empirical observation that became self-fulfilling after some adjustments.

Ten years later, in 1975, Gordon Moore revised "Moore's Law", saying that the doubling of transistors per chip was now occurring every two years, instead of every year. Then, in 1997, Gordon Moore revised "Moore's Law" once again, showing that the doubling of transistors was now occurring every 18 months. These repeated revisions affirm that "Moore's Law" was not actually a law of nature but an interesting, if temporary, phenomenon.

In science and engineering, we have laws that predict outcomes when variables change, like the first and second laws of thermodynamics, Newton's laws of motion or Maxwell's equations. They don't change over time, unlike Moore's Law (Figure 2). Even Dr. Moore pointed out, in his ISSCC keynote in 2003, that "no exponential is forever".

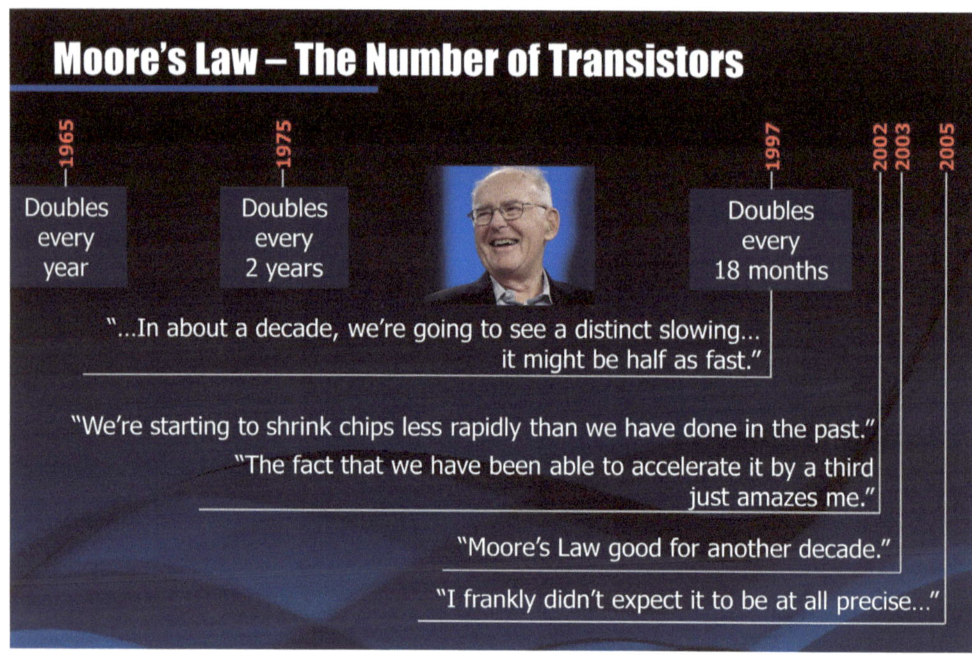

Figure 2. Moore's "Law" evolved over time.

Why did "Moore's Law" take on such significance and work so well, despite the adjustments in time scale? The answer is that "Moore's Law" is based upon an actual law of nature called the "learning curve" (See Figure 1 in Chapter 1). Learning curves have been used over the last hundred years to predict the future cost per unit of products as diverse as airplanes, beer and transistors. They were used strategically by Texas Instruments in the 1960s to "forward price" new semiconductor components in order to achieve a desired future market share and profitability.

The learning curve and Moore's Law are actually the same when two conditions are met. These are:1) If most of the cost reduction for semiconductor chips comes from shrinking feature sizes and growing wafer diameters and 2) If the cumulative number of transistors manufactured by the semiconductor industry increases exponentially with time. If these two conditions are met, then Moore's Law and the learning curve become straight lines that predict the same trend (Figure 3).

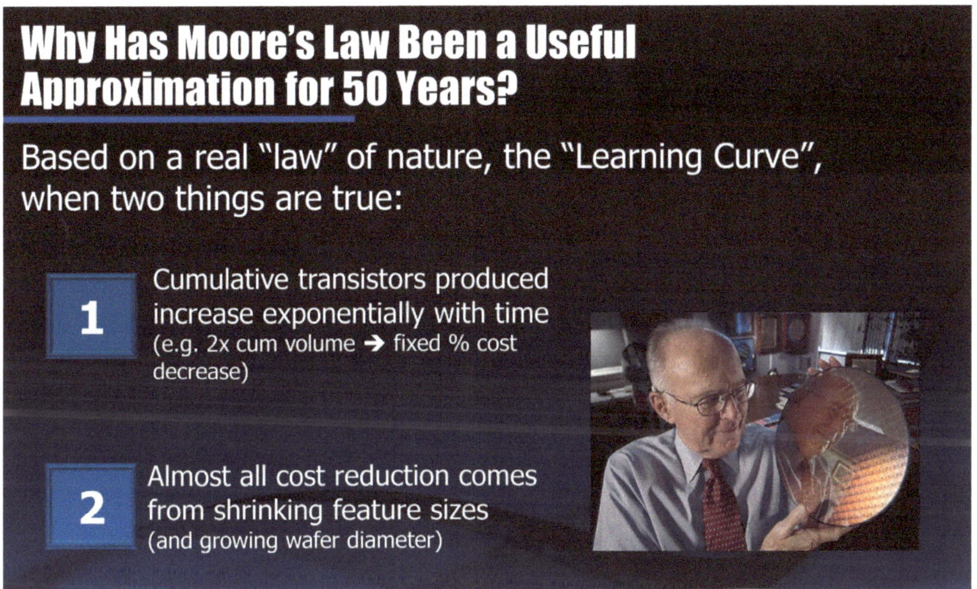

Figure 3. Learning Curve and Moore's Law are the same under certain conditions.

If "Moore's Law" is based upon a real law of nature, i.e. the learning curve, then why did it have to be adjusted from one year to two years and then back to eighteen months? The answer comes from assumption number two above and is shown in Figure 4. Even though the number of transistors shipped each year has grown exponentially through most of the history of the semiconductor industry, there was a period when growth slowed and then later returned to the exponential trend.

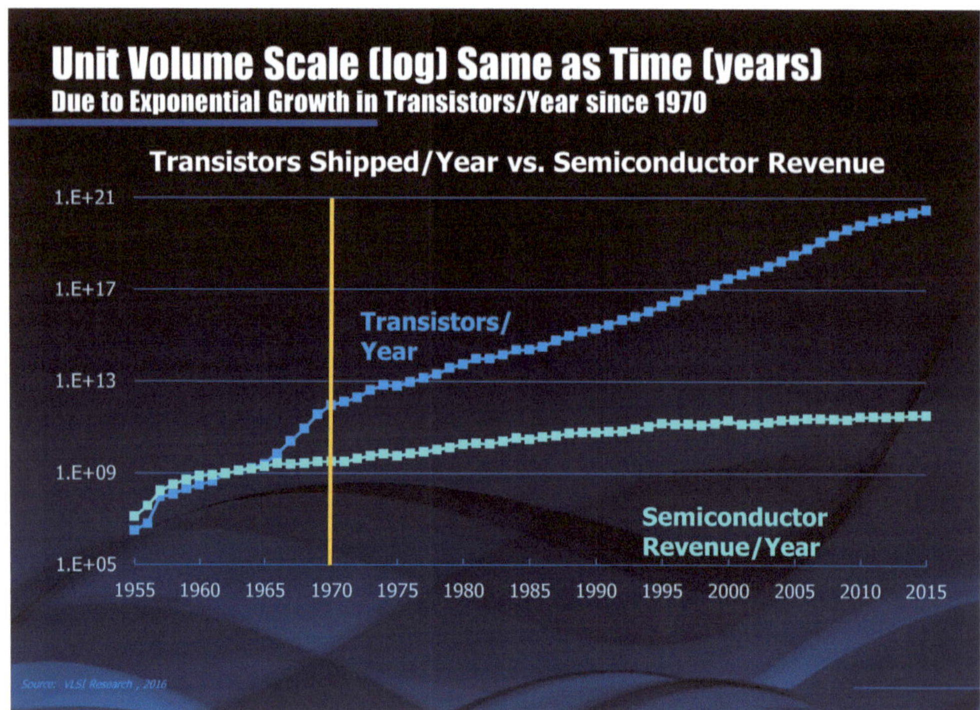

Figure 4. Growth in cumulative number of transistors has not always been exponential with time.

That change in growth rate caused "Moore's Law" to increase from one to two years and then back to eighteen months. Because the learning curve is a log/log graph, exponential growth of the cumulative number of transistors produces a straight line with time as well as with the cumulative number of transistors. Unlike Moore's Law, the learning curve works well even if the exponential growth of units deviates. Moore's Law uses time as its horizontal axis so linearity is assured only if cumulative transistor growth is exponential.

Today, many people worry that the inevitable end of Moore's Law will leave us with a stagnant semiconductor industry with no guideposts to drive new silicon technology directions. Fortunately, these people need not worry. The learning curve is valid forever (when measured in constant currency, corrected for governmentally-induced inflation) as long as free market economics prevail, i.e. negligible trade barriers, no regulatory price controls, etc.

Figure 5 shows a learning curve for the electronic switch measured as revenue per MIP, beginning with vacuum tubes and progressing through germanium and quickly transitioning to silicon.

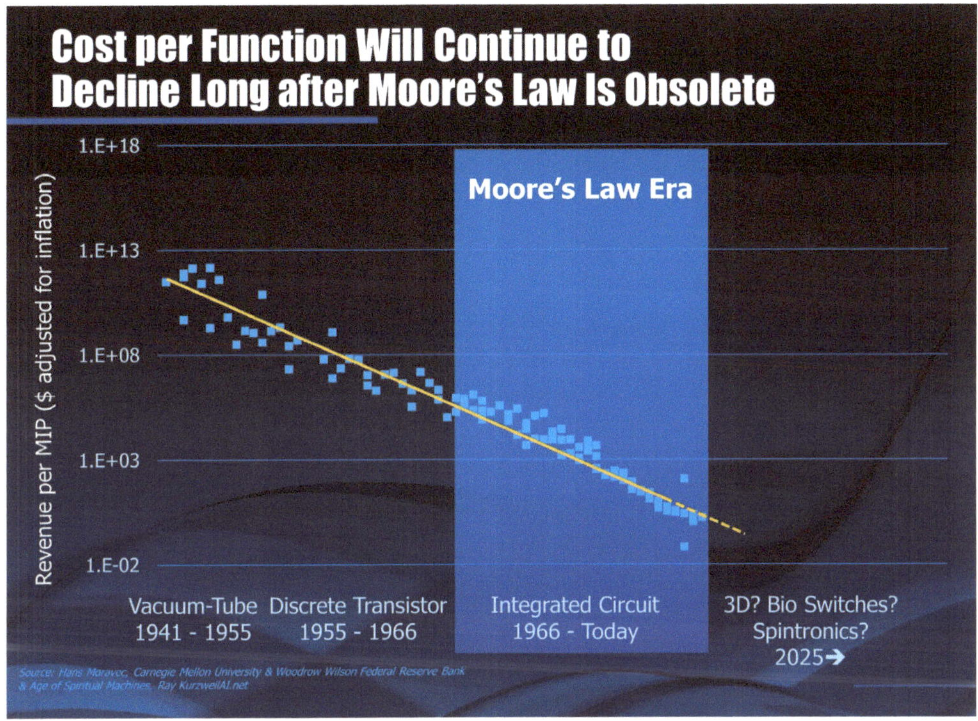

Figure 5. Cost per Function, or per MIP, transcends the transistor era.

We use industry revenue for the vertical axis, instead of cost, because the data is more readily available but the two variables should be surrogates for one another, as long as gross margin percentage doesn't vary much. The horizontal axis is the cumulative number of switches shipped throughout history. That number has been available from the Semiconductor Industry Association, as well as from other semiconductor analysts, for decades.

Of course, the learning curve for electronic switches doesn't care whether the cost reduction is achieved with mechanical switches, vacuum tubes, transistors or even carbon nanotubes in the future. The learning curve is technology independent if a more generalized unit than transistors is measured. We therefore have a metric to track when the further improvements in cost or power are so difficult with silicon that we have to consider an alternative like carbon nanotubes or bio-switches.

The important result of this information for the electronics industry is that the death of Moore's Law doesn't lead to random, unpredictable trends in semiconductor technology. We have a road map. As long as we can measure the growth rate of transistor shipments, we will know the cost or revenue per transistor of the semiconductor industry, or vice versa.

Chapter 4: Gompertz Predicts the Future

In 1825, Benjamin Gompertz proposed a mathematical model for time series that looks like an "S-curve".[1] Mathematically it is a double exponential (Figure 1) where y=a(exp(b(exp(-ct)))) where t is time and a, b and c are adjustable coefficients that modulate the steepness of the S-Curve. The Gompertz Curve has been used for a wide variety of time dependent models including the growth of tumors, population growth and financial market evolution.

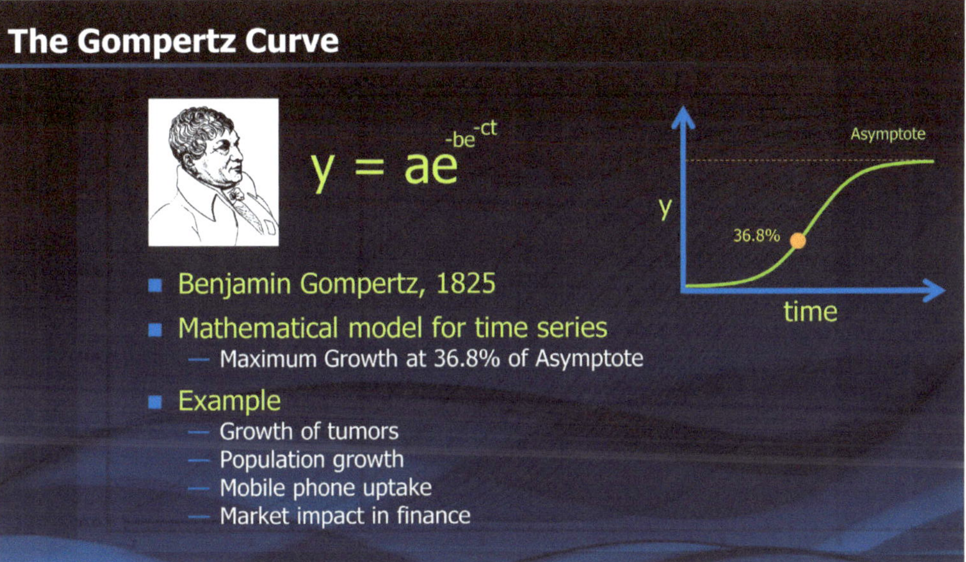

Figure 1. The Gompertz Curve.

S-Curves are common in nature. In any new business, or in biological phenomena, we start out small with an embryonic business or a tiny cell and it reproduces slowly but the percentage growth rate is large. As time goes on, the growth accelerates until it finally slows down as it reaches saturation. A new product takes a significant period of time for early adopters to spread the word of its benefits but it then goes viral, saturates the market and then declines (Figure 2).

On the right half of Figure 2, we see the same phenomena when the vertical axis of the graph is the cumulative number. An example would be the freezing of water in a pond. It starts with a few water molecules and then grows

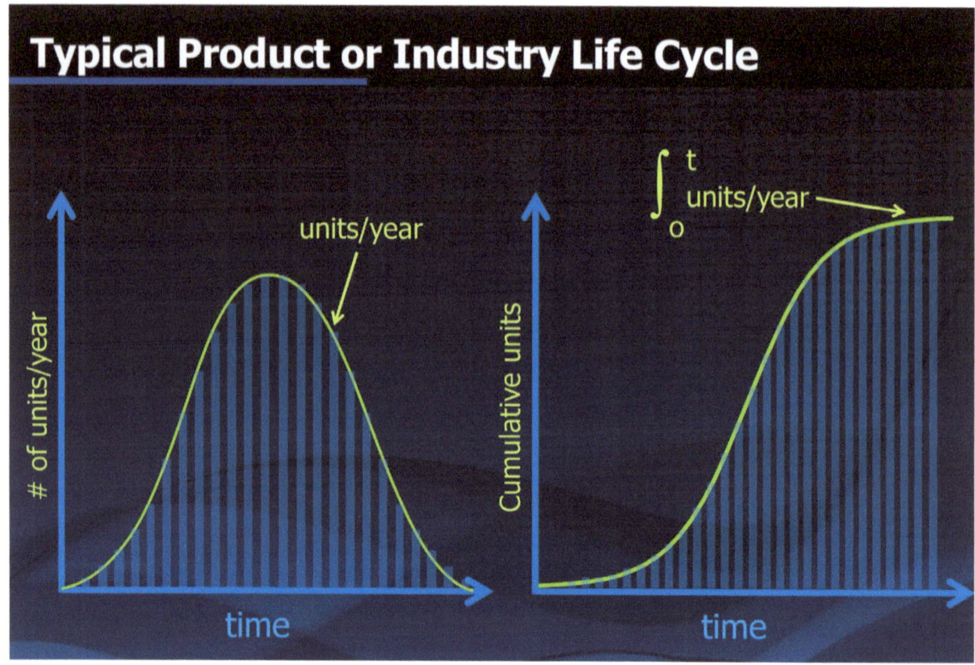

Figure 2. Typical product life cycle or life cycle of an industry.

to a critical nucleus which grows rapidly until the pond is mostly frozen. Then the last bit of water freezes over a longer period of time. Expressed mathematically, the integral of the cumulative function is the area under the curve and it increases until the S-Curve finally flattens.

Figure 3 shows the stages of growth of the S-Curve. It starts out slow but the highest percentage growth is early in the S-Curve evolution. The curvature of the "S" increases upward until about 37% of the time on the horizontal axis is completed.[2] Then the curvature is downward. Mathematically we would say that the second derivative of the Gompertz function is positive until about 37% of the time is completed and then the second derivative becomes zero. The rate of the rate of growth becomes negative and so the growth rate is less each year after that point.

I first became acquainted with the Gompertz Curve while managing a design project that TI was doing for IBM. IBM wanted us to report the number of simulated transistors that we had completed in our design each week. They then plotted them as a Gompertz Curve (Figure 4).

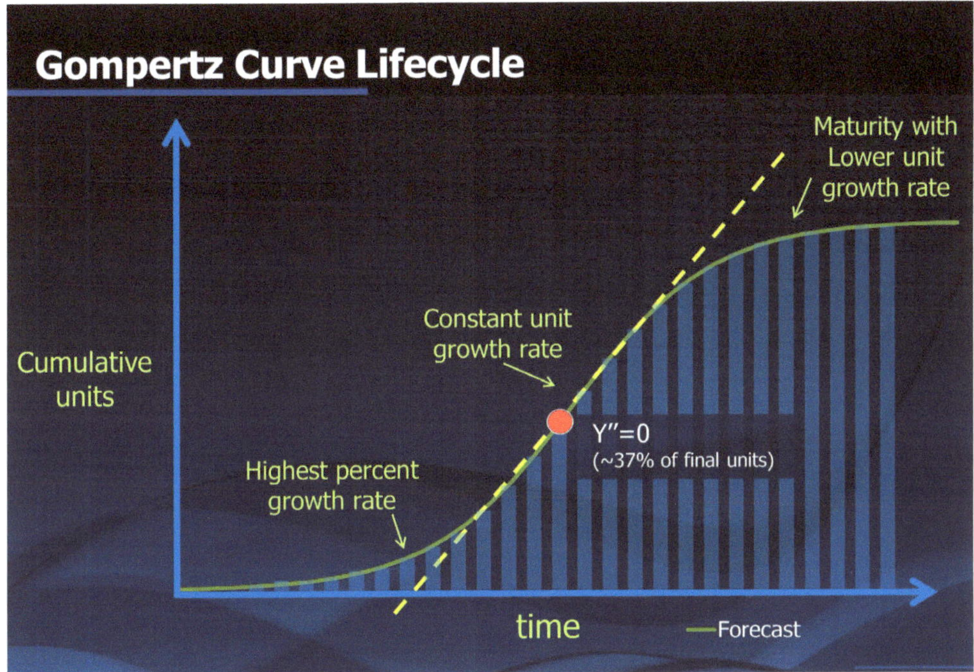

Figure 3. Gompertz Curve Life Cycle.

Inexperienced project managers would have been frustrated by the fact that progress was initially very slow. The specification for the design project kept changing, new architectural approaches were tested and the number of simulated transistors remained small for some time. Then, things took off.

The number of transistors completed each week grew linearly. Our inexperienced design manager would have been delighted and would have extrapolated this progress to an early completion as shown in Figure 4. With more experience, he would realize that the last fifth of the project would take more than one third of the total time.

While the Gompertz Curve is useful for project management, it provides even more insight for forecasting the future success of an embryonic product. Figure 5 shows the evolution of worldwide sales of notebook PCs. Using the data available to us with the actual shipments of PC notebooks in the years up through 2001, we can solve for the Gompertz coefficients a, b and c. We could then have used these coefficients to predict the future evolution of the growth curve for cumulative units of PC notebooks shipped.

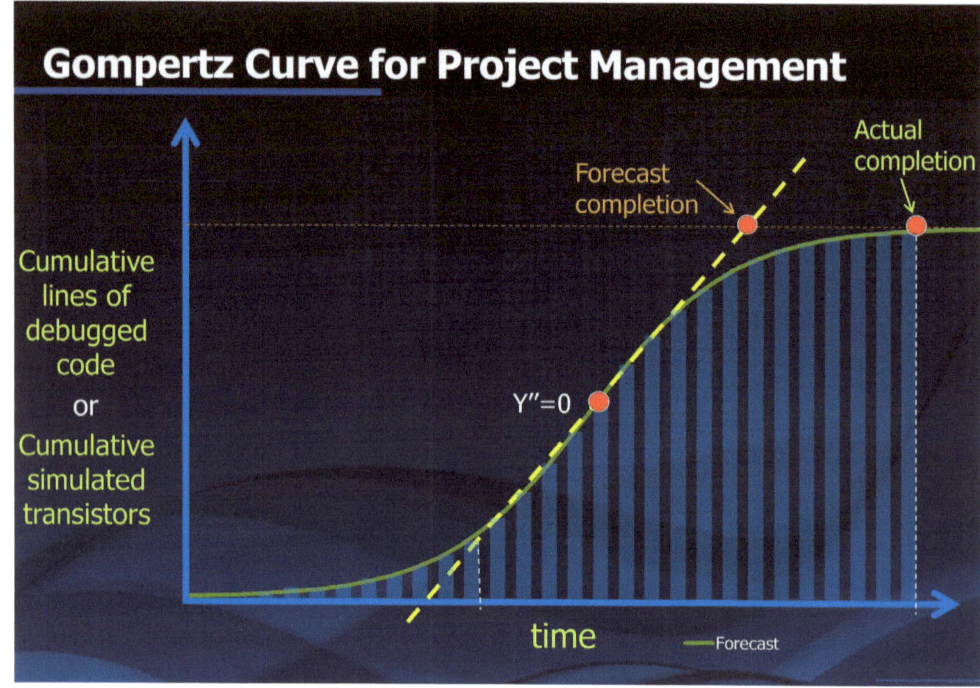

Figure 4. Use of Gompertz Curve for Project Management.

Figure 6 shows the Gompertz prediction versus the actual results reported in 2016. The results are nearly identical.

If you were an aspiring competitor in the PC notebook business in 2001, or even an investor in the personal computer business, accurate knowledge of the future market for PC notebooks over the next fifteen years could be very useful.

Finally, Gompertz Curves can be used to predict the future of an industry. A good choice would be the future of the silicon transistor since lots of research dollars have been devoted to developing an alternative to the silicon switch and we don't even know how soon we need it. Or do we? Gompertz analysis provides an opinion. It's shown in Figure 7.

Although the semiconductor industry and silicon technology may seem mature to some, we are in the infancy of our production of silicon transistors. The cumulative number of silicon transistors produced thus far is almost negligible compared to the future, as shown in Figure 7.

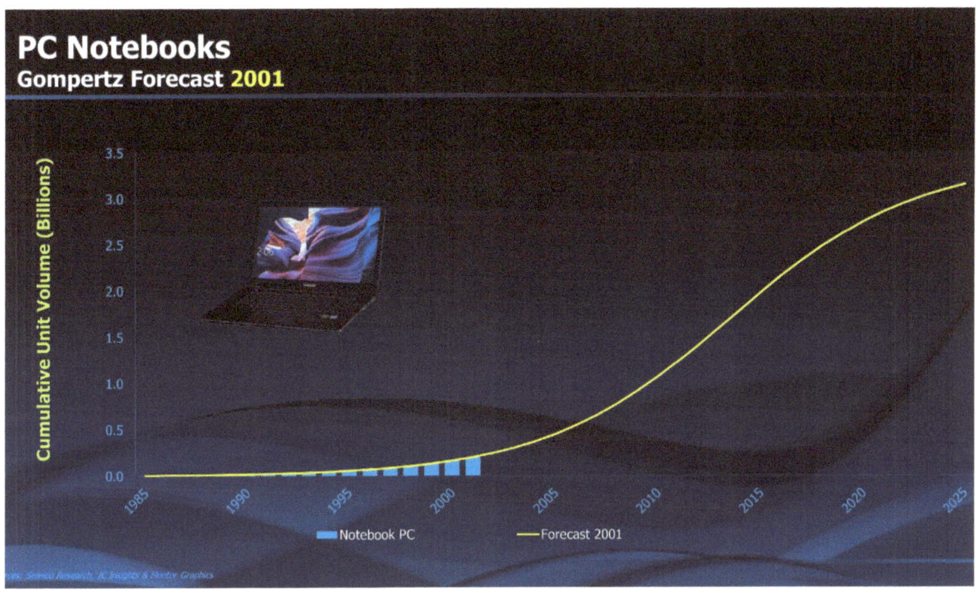

Figure 5. PC Notebook Shipments through 2001 provide data for Gompertz forecast.

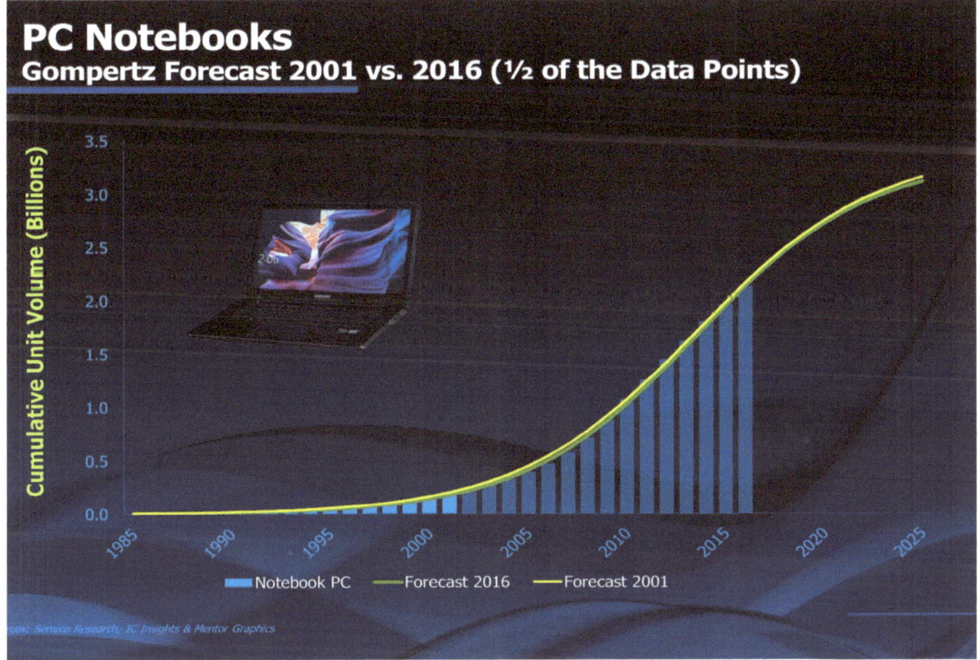

Figure 6. Actual PC Notebook shipments though 2016 (shown in green) versus Gompertz prediction in 2001 (shown In yellow).

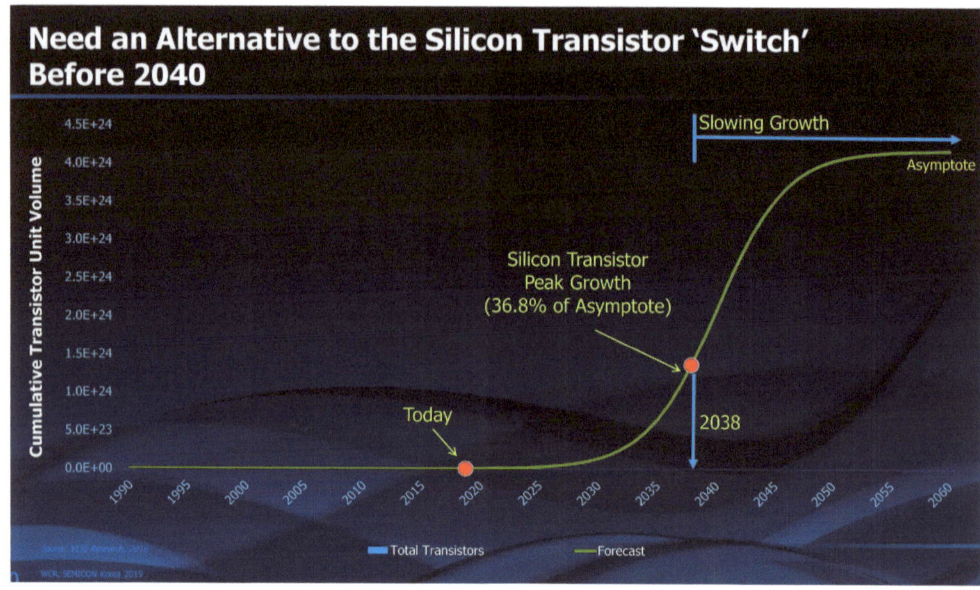

Figure 7. Future of the silicon transistor.

The actual RATE of growth of shipments of silicon transistors is predicted to increase until about 2038. At that time, the Gompertz Curve suggests that the increase in the RATE of growth will become zero and the RATE of increase will be less each year until we reach saturation, sometime in the 2050 or 2060 timeframe. By then, we should have developed lots of alternatives.

1 https://en. wikipedia. org/wiki/Benjamin_Gompertz

2 https://arxiv. org/ftp/arxiv/papers/1306/1306. 3395. pdf

Competitive Dynamics in the Semiconductor Industry

Chapter 5: Consolidation of the Semiconductor Industry

For the last decade, semiconductor industry analysts have been writing articles and giving presentations that predict the increasing consolidation of the industry to the point where a few large companies dominate worldwide sales of semiconductor components. In recent years there has been some justification for this view as the combined market share of the top five companies in the industry has increased, as has the combined market share of the top ten.

The general thesis of these discussions of semiconductor industry consolidation is the widely accepted model of growth and maturation of an industry. Industries like steel, automobiles and others that have propelled decades of economic expansion in the world should grow rapidly in their youth and then slow down as their markets saturate and stabilize. During this period approaching maturation, revenue growth is not large enough to drive increased profit and enterprise value so the focus becomes cost reduction. Figure 1 shows the consolidation of the U.S. steel industry.

Figure 1. Steel industry consolidation in the U. S. [1,2]

By becoming more efficient, these mature industries reduce their labor and material costs, acquire competitors to achieve better economies of scale and reduce their research and development expenses since their industry is no longer evolving rapidly and there are fewer opportunities for new product and technology innovations. The acquisition process eventually leads to an oligopoly of a few large surviving companies that can achieve the required economies of scale to prosper despite their slow or declining revenue.

There are at least two problems with this kind of analysis. First, the assumption that industries mature and consolidate down to a few large enterprises may be the exception rather than the rule. Second, the analysis of the semiconductor industry as a candidate for this model is probably premature since we're seeing new growth in revenue and profits and innovation, at the time of this presentation in 2016, despite the sixty year age of the semiconductor electronics industry.

Consider first the assumption that most industries eventually consolidate. While consolidation certainly occurred in the U. S. steel industry in the 1960s and employment has now been reduced by nearly 85%, the number of steel companies was only reduced by 50%.

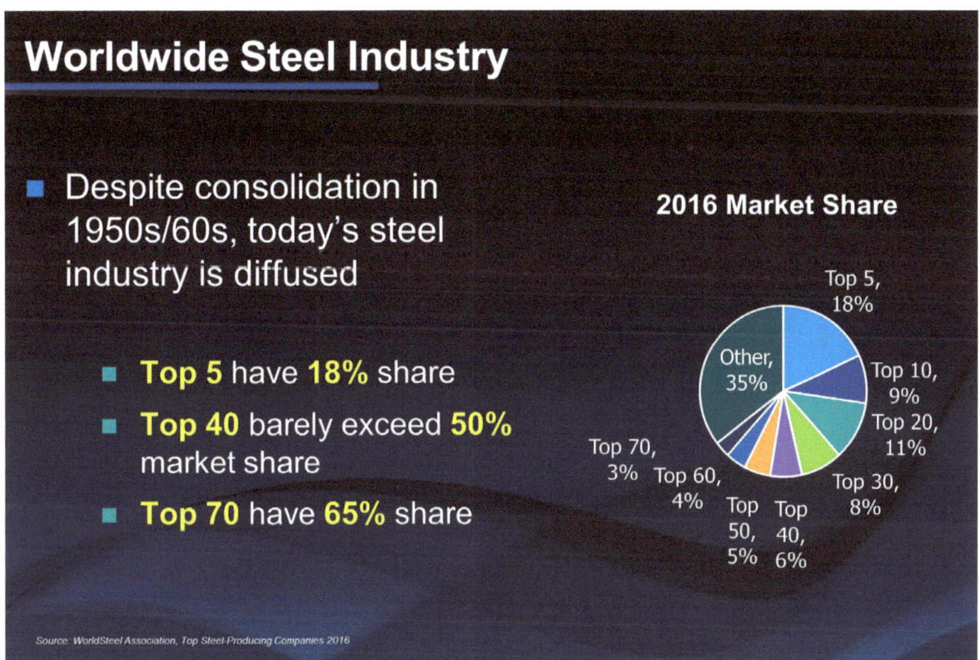

Figure 2. Competitive state of the worldwide steel industry.

New technology provided by mini mills created a set of new competitors in the industry. Worldwide, consolidation of the steel industry has left us with far more than the classical oligopoly of companies (Figure 2). The five largest steel companies in the world account for only 18% of the revenue of the industry and it takes forty companies to account for half of the worldwide steel production.

The case of the automobile industry, though different, also provides insight into the maturation process of industries. Figure 3 shows the growth of the automotive industry, reaching a peak of 272 companies in 1909 and consolidating down to GM, Ford and Chrysler with 91% U. S. market share in the 1960s.

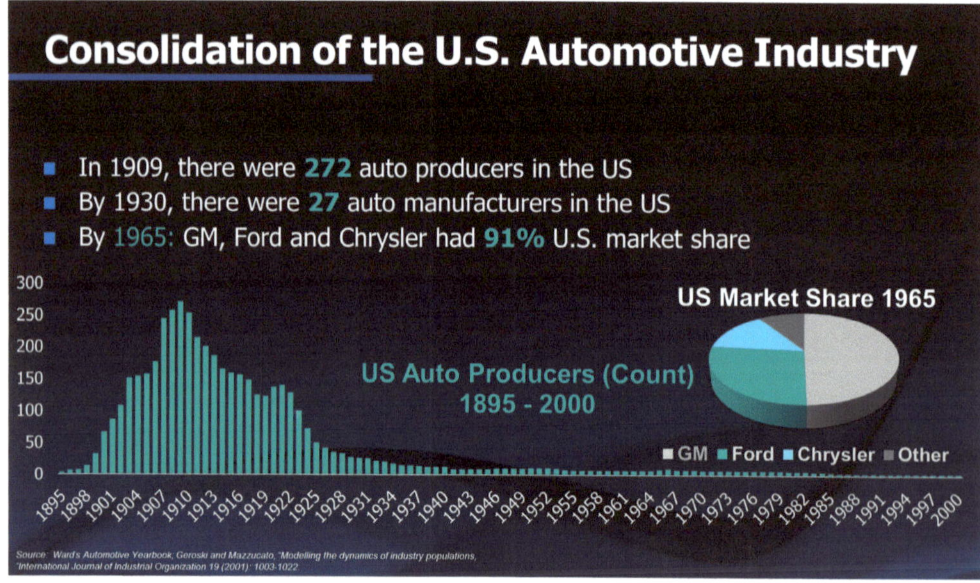

Figure 3. Growth of the automobile industry.

This oligopoly was temporary, however, as foreign manufacturers from Europe, Japan and Korea gained market share in the U. S., passing the combined market share of GM, Ford and Chrysler in 2007. Emergence of electric cars and evolution of technology for driverless cars has stimulated the emergence of nearly 500 new companies that have announced plans to produce electric cars and light trucks in the near future and nearly 200 planning driverless cars.

Are there any industries that consolidate down to an oligopoly and remain that way? The answer is, "yes, but…. ". The well accepted model of consolidation seems to work in industries that operate in relatively free worldwide markets that are largely free of regulatory and tariff barriers and have a low cost of transport so that products can flow easily from one region to another. Two examples of this are the hard disk drive and the dynamic RAM (dynamic random-access memory) businesses.

The number of competitor companies in the hard disk drive industry peaked at 85. Figure 4 shows the current state of that industry with three participants controlling almost 100% of the revenue of the industry.

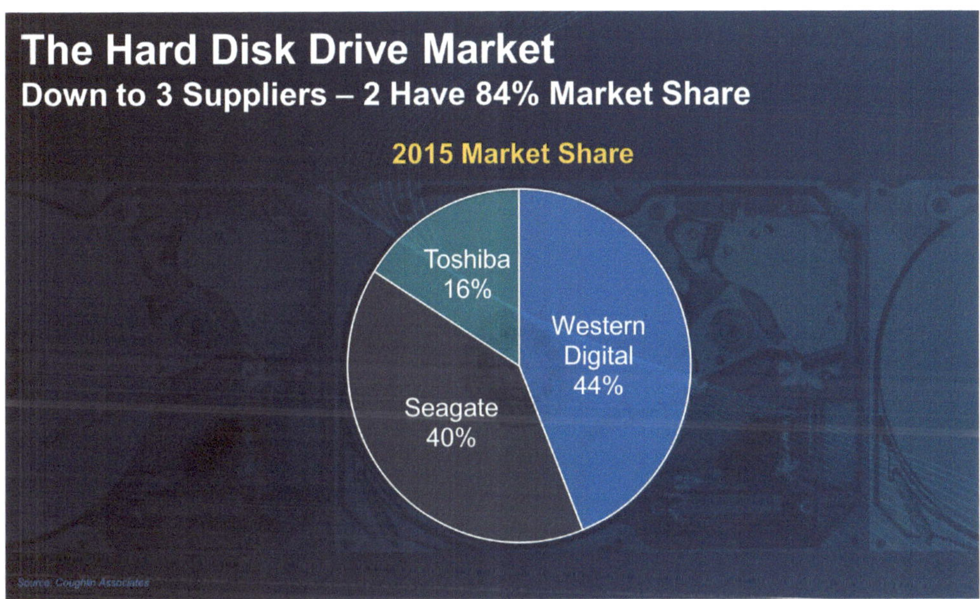

Figure 4. *Market shares of the leading hard disk drive manufacturers in 2017.*

But like most industries, technical discontinuities change the game. Emergence of solid-state storage to replace rotating media hard disk drives is changing the market share outlook (Figure 5). Samsung is emerging as the new leader partly because of its leading position in the NAND FLASH component business.

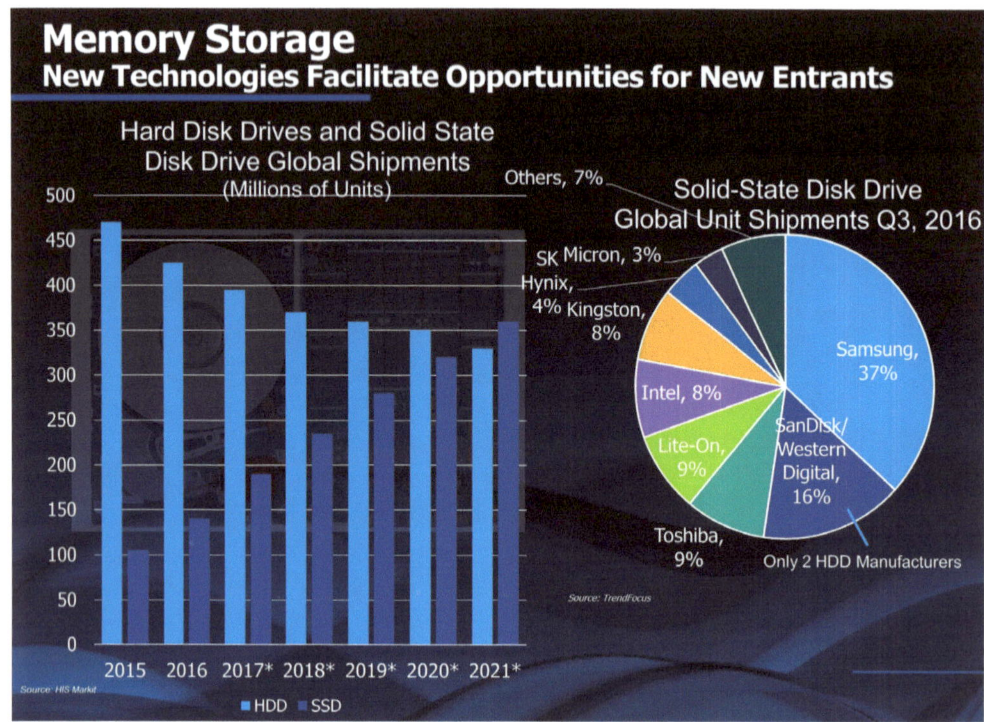

Figure 5. Solid state storage changes the competitive landscape.

The other example of the consolidation of an industry is the DRAM business (Figure 6). In 1997, the top three producers of dynamic RAM had less than 40% of the market. By 2014, they had 95%. Both DRAMs and hard disk drives satisfy the requirement of low cost of transport. They are also industries that have relatively free market design, production and distribution worldwide.

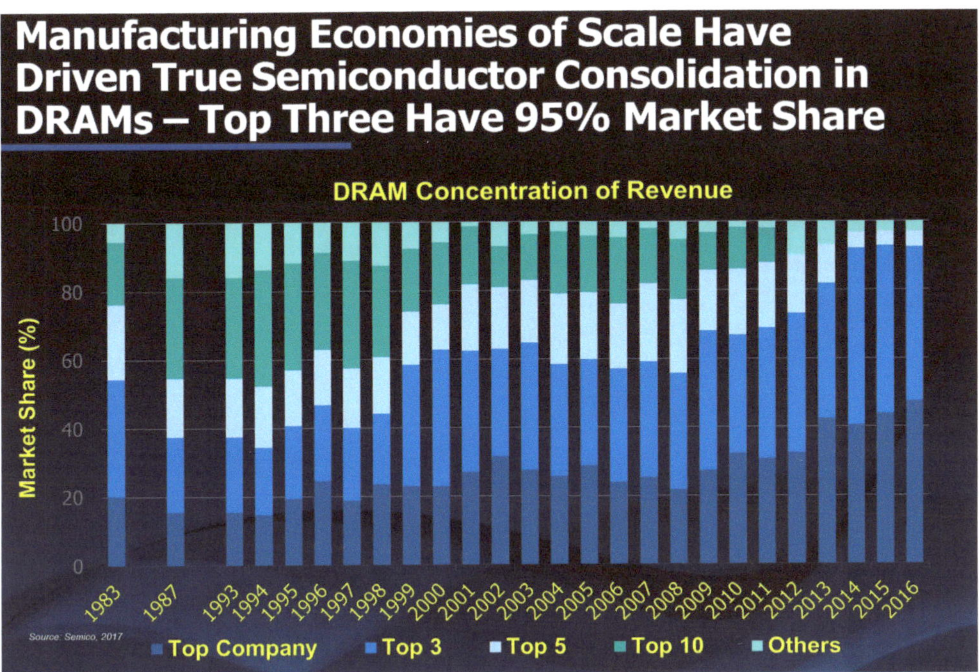

Figure 6. DRAM worldwide market share. Combined share of the three largest companies grew from about 35% in 1994 to 95+% in 2016.

How does all this relate to the broader semiconductor industry? Will it consolidate down to a dominant few companies and remain there, as the analysts suggest? It's doubtful, at least for the near term. Let's look at the history of semiconductor industry consolidation, or more accurately, its "deconsolidation".

Since 1965, the semiconductor industry has been "deconsolidating" (Figure 7). In 1966, three companies, TI, Fairchild and Motorola, shared about 70% of the total semiconductor market. Over the next seven years, that share dropped to 53%, driven by new entrants like National Semiconductor, Intel, AMD, LSI Logic and about 25 more.

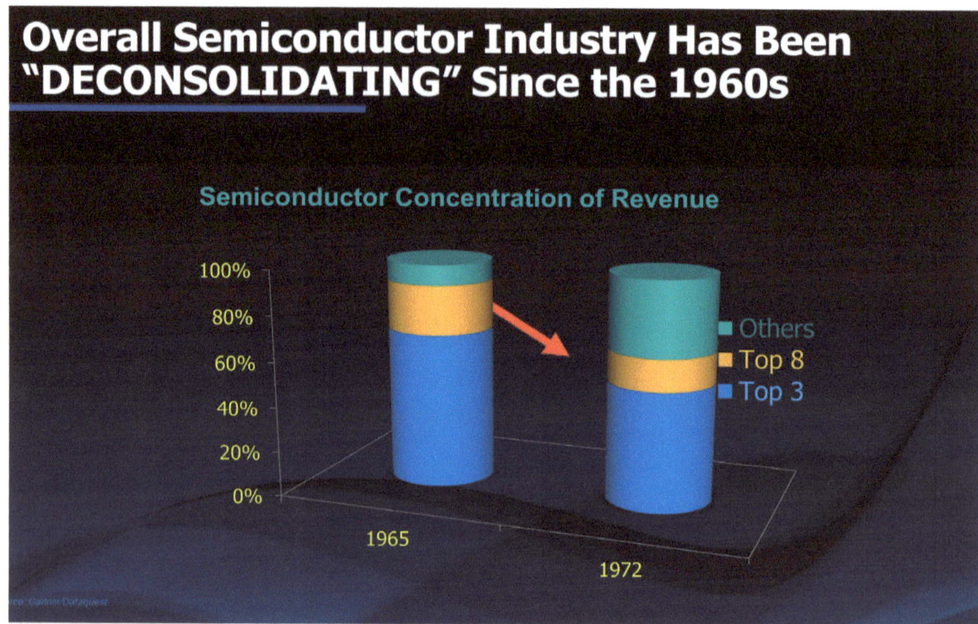

Figure 7. Semiconductor industry deconsolidation from 1965 to 1972.

Over the next 40 years, the market share of the top semiconductor company remained roughly the same, at 12% to 15% market share, although the names changed from TI in 1972 to NEC and then to Intel. Combined market shares of the top five and top ten semiconductor companies decreased or remained flat during this period (Figure 8).

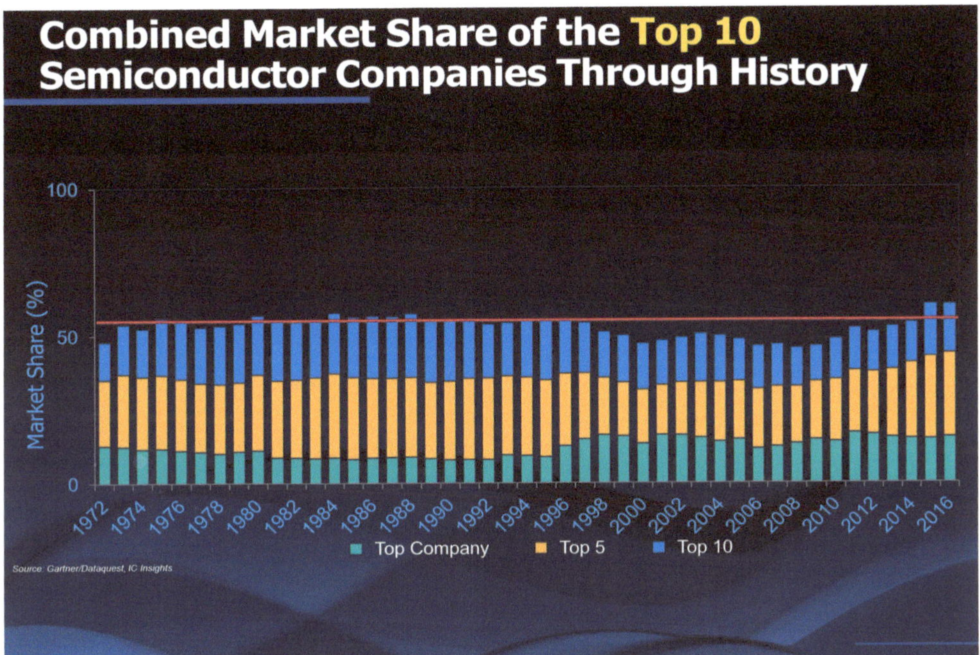

Figure 8. Combined market share of the five and ten largest semiconductor companies.

During 2016 through 2018, the combined market share of the top ten semiconductor companies increased modestly, partly due to an unusual increase in DRAM unit prices as well as a very strong computer server market that favored Intel. The most remarkable piece of data is shown in Figure 9. Throughout history, the combined market share of the fifty largest semiconductor companies has been decreasing.

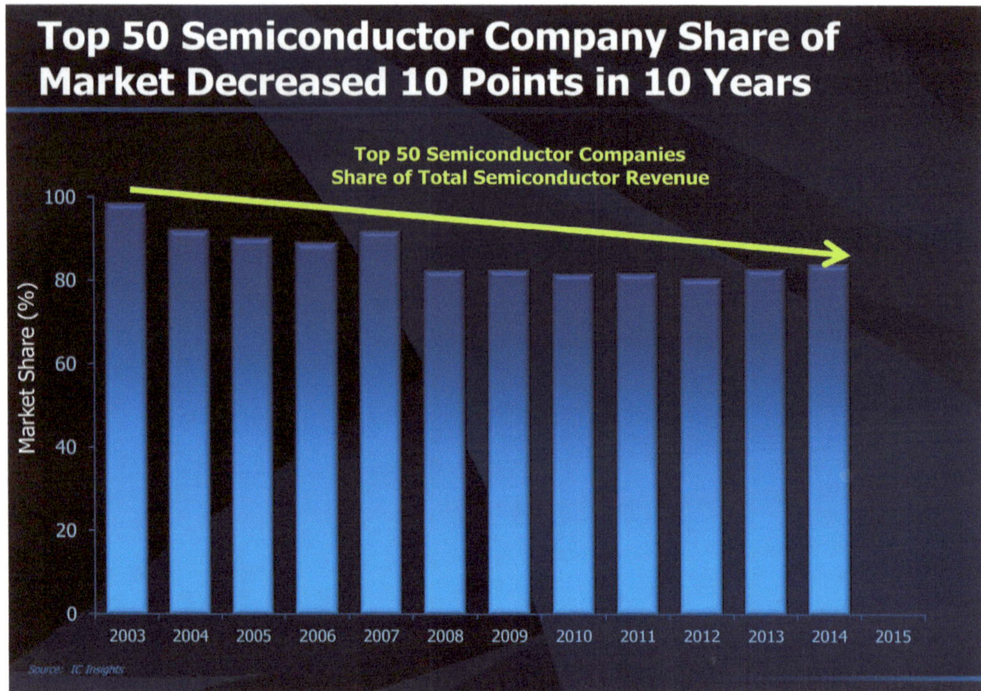

Figure 9. Combined market share of the fifty largest semiconductor companies from 2003 through 2014.

This observation says a lot about the character of the semiconductor industry both now and throughout history. Company leadership in the industry is continuously changing as new technologies emerge and new companies secure the leading market share in these new technologies.

Figure 10 shows the top ten ranking of semiconductor companies over a fifty year period. The company names shown in green are ones that have dropped out of the top ten and never reappeared except for NXP.

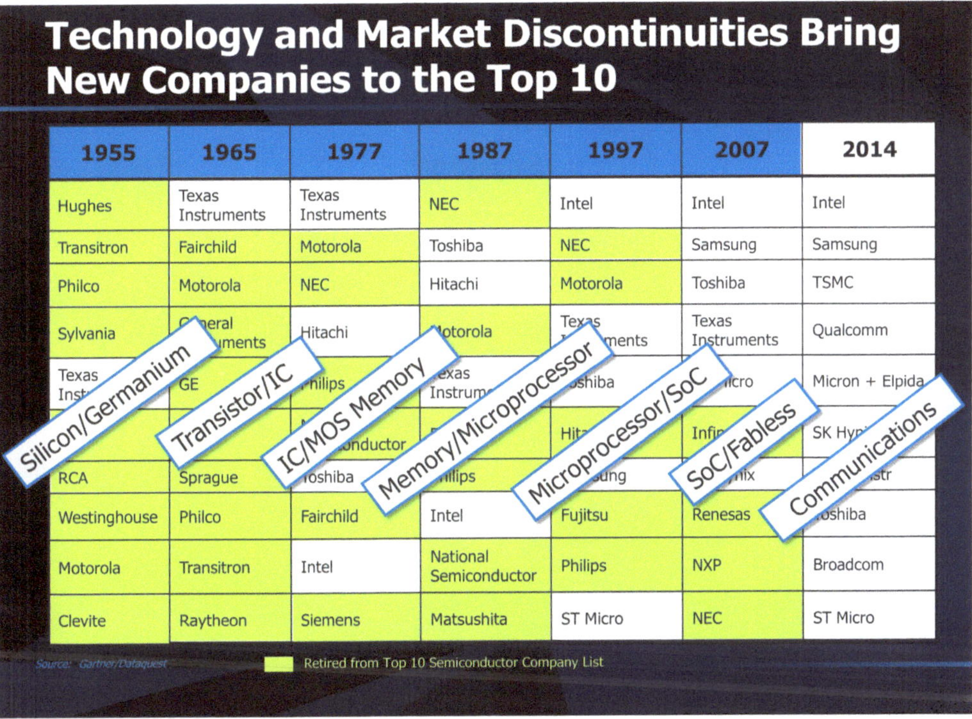

Technology and Market Discontinuities Bring New Companies to the Top 10

1955	1965	1977	1987	1997	2007	2014
Hughes	Texas Instruments	Texas Instruments	NEC	Intel	Intel	Intel
Transitron	Fairchild	Motorola	Toshiba	NEC	Samsung	Samsung
Philco	Motorola	NEC	Hitachi	Motorola	Toshiba	TSMC
Sylvania	General Instruments	Hitachi	Motorola	Texas Instruments	Texas Instruments	Qualcomm
Texas Instruments	GE	Philips	Texas Instruments	Toshiba	Micron	Micron + Elpida
[illegible]	[illegible]	National Semiconductor	[illegible]	Hitachi	Infineon	SK Hynix
RCA	Sprague	Toshiba	Philips	Samsung	Hynix	Texas Instr
Westinghouse	Philco	Fairchild	Intel	Fujitsu	Renesas	Toshiba
Motorola	Transitron	Intel	National Semiconductor	Philips	NXP	Broadcom
Clevite	Raytheon	Siemens	Matsushita	ST Micro	NEC	ST Micro

Source: Gartner/Dataquest

Retired from Top 10 Semiconductor Company List

Figure 10. Top ten semiconductor companies change with time. Companies shown in green fell out of the top ten.

The number of companies that have retired from the top ten is greater than half of all those who have ever been in the top ten. Only Texas Instruments has remained in the top ten throughout the fifty year period and even it is probably destined to drop out as it focuses its business in analog and power and further disengages from the high volume "big digital" chips that constitute, along with memory, so much of the semiconductor revenue today.

It's difficult for semiconductor companies to reinvent themselves as new growth markets emerge. The large semiconductor companies tend to grow at about the overall semiconductor market average growth rate while the new entrants grow much faster, albeit from a smaller revenue base. Gradually, these small companies climb the ranks on their way to the top ten.

Will the wave of merger mania in 2016 and 2017 continue into the future as the semiconductor industry finally matures and consolidates? Surely the competitive advantage of scale will lead to more mergers and a more difficult environment for small companies to compete without the scale of the big ones? The recent slowing of merger activity, although significantly affected by government regulatory disapprovals, suggests that we may not have reached that stage of consolidation (Figure 11).

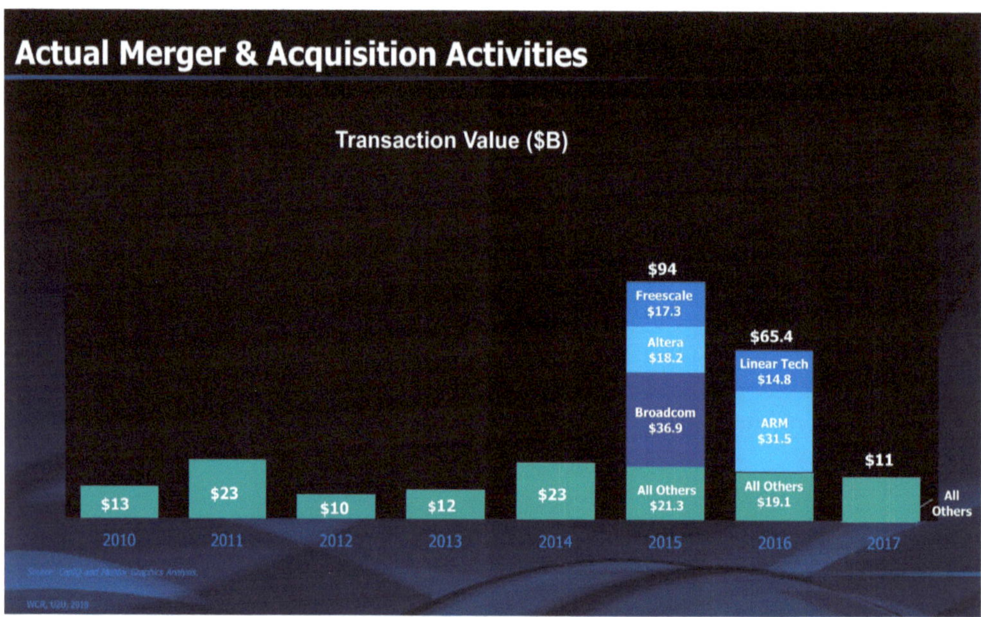

Figure 11. Value of semiconductor industry mergers by year.

Actual numbers make 2017 and 2018 among the lowest dollar value of major merger years in recent history, both in number and in enterprise value. The recent increase in semiconductor industry revenue growth rate to 22% in 2017 after two years of no growth also suggests that the announcement of industry maturity may have been premature.

In the next chapter, we will examine the factors behind the consolidation that has been occurring. A reasonable conclusion would be that the limited amount of consolidation that is occurring in the semiconductor industry is not motivated by size or broad economies of scale but by specialization. Profitability in the semiconductor industry is driven by market share in very specific specialties and the industry is in a transition to increased specialization which is also increasing overall profitability.

1 https://www. nwitimes. com/business/local/steel-ceo-more-consolidation-inevitable/article_c407cc83-7d1b-59eb-a838-f1ea1723845c. html

2 https://247wallst. com/investing/2010/09/21/americas-biggest-companies-then-and-now-1955-to-2010/

Chapter 6: Specialization in the Semiconductor Industry

Recently, the combined market share of the top ten and top twenty semiconductor companies has been increasing, contrary to the trend of the last fifty years. Given the acceleration in mergers and acquisitions that began in 2015, one might assume that, as the semiconductor industry approaches maturity, companies are consolidating to increase their competitive advantage through economies of scale. After all, that's what many industries, including disk drives and DRAM's have done in the past.

Closer examination of this trend, however, indicates that semiconductor companies are moving toward specialization rather than just bulking up to increase their revenue. Let's look at the top five largest semiconductor companies, where the consolidation is most evident. The combined market share of these companies has been increasing in recent years as they grow at a 9% compound average growth rate (CAGR) versus a market that grew at 2% CAGR through 2017 (Figure 1).

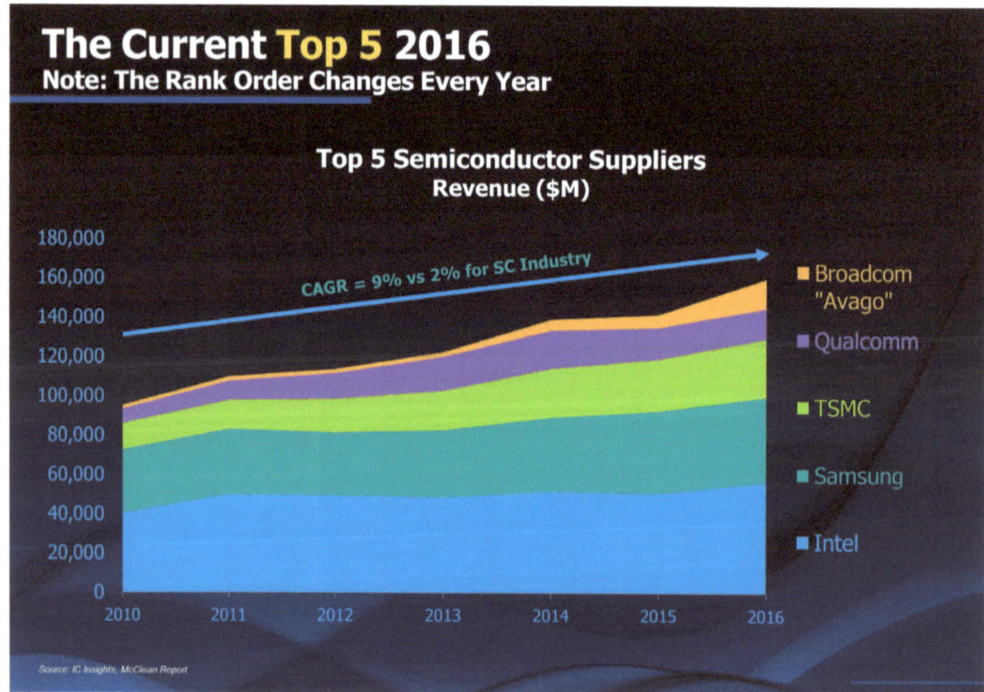

Figure 1. Increasing combined market share of the five largest semiconductor companies.

Did they grow by acquisition of other companies? In general, "no".

Despite acquisitions like Altera, Intel's market share over the period from 2010 to 2016 was flat at about 15.5%. Samsung gained market share during the period, moving from 10.2 to 12.1% but this gain was not caused by acquisitions. TSMC, the third largest semiconductor company by revenue, grew its market share substantially during the period, rising from 4.5 to 8.1% with no acquisitions. And Qualcomm's gain in market share from 3.1 to 4.2% was almost totally driven by the growth of its primary market, wireless telecommunications, rather than any acquisitions. Only Broadcom grew by acquisitions during the period, moving from 0.7 to 4.2% market share.

There were indeed companies that grew economies of scale through acquisitions during the period 2010 through 2016 but they are not a significant share of semiconductor industry revenue. They include the TriQuint/RFMD merger to form QORVO, International Rectifier/Infineon, On Semiconductor/ Fairchild, and Linear Technology/Analog Devices, to name some examples.

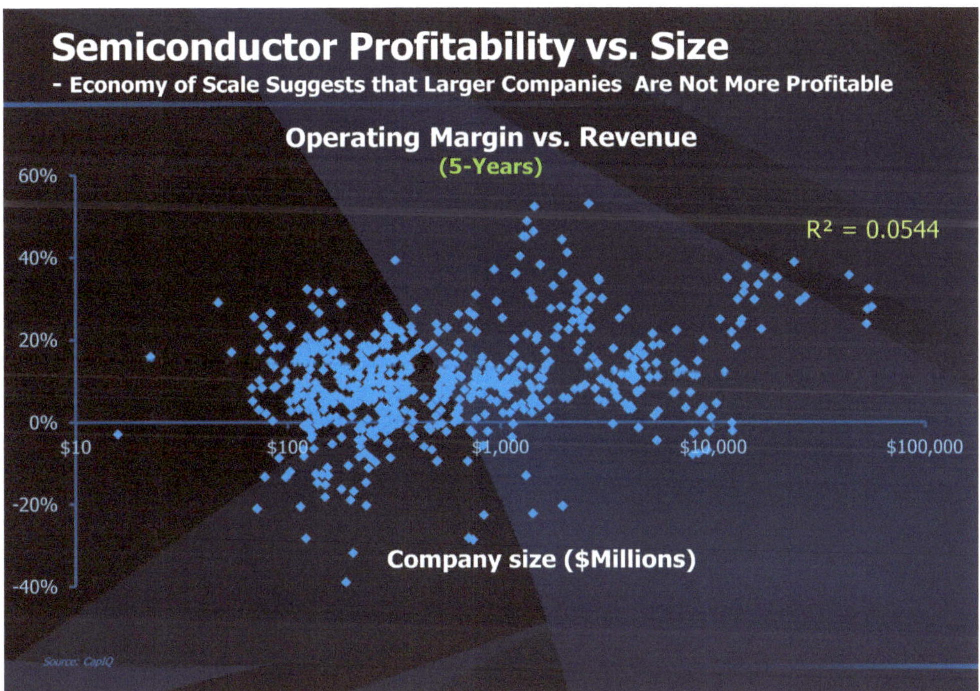

Figure 2. Lack of correlation between semiconductor revenue and operating profit percent of the largest semiconductor companies 2010 through 2016.

Overall data for the industry suggest that there is no correlation between operating profit percent and revenue, with a correlation coefficient of only 0.0706 (Figure 2).

Why then is there an accelerated level of semiconductor mergers and acquisitions in 2015 and 2016? It turns out that companies that used acquisitions and divestitures to specialize their businesses usually improved operating profit percent more than those who did not. Texas Instruments is a good example (Figure 3).

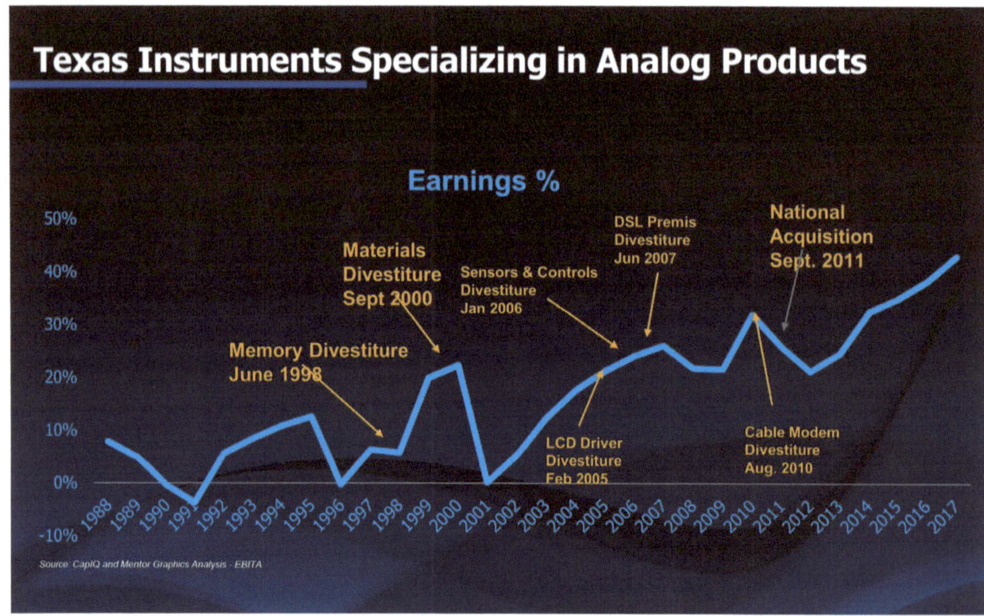

Figure 3. *Texas Instruments operating profit percent.*

When I worked at TI in the 1970s and 1980s, the company made almost every conceivable type of semiconductor component. One could say that TI made everything in the semiconductor business except money. Through a series of acquisitions, divestitures and business terminations since the year 2000, TI has focused its business on analog and power components. As a result, TI has progressed from profitability that averaged less than 10% operating profit to a 40% operating profit in 2017, the highest of the major companies in the semiconductor industry.

NXP is another good example (Figure 4). In 2014, nearly 30% of its revenue came from "standard products". Over the next five years, this percentage became negligible and more than 90% of NXP's revenue then came from two major areas, automotive and security.

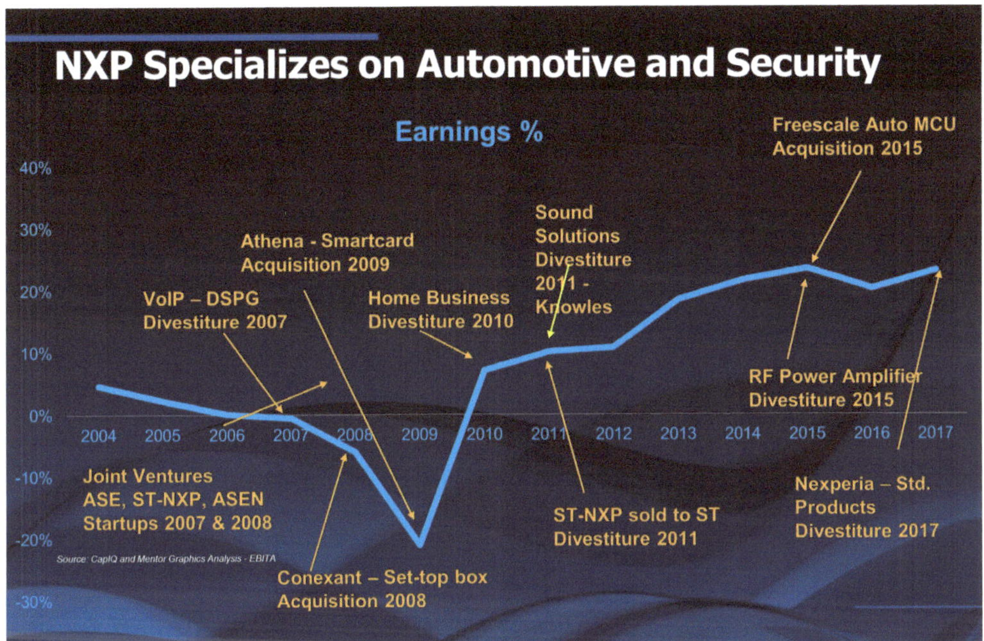

Figure 4. NXP Operating margin after adjustment for extraordinary items.

AVAGO is a similar story although the specialization was achieved by an aggressive series of acquisitions (Figure 5). Along with the acquisitions came divestitures resulting in very strong market share in wireless communications and networking, a specialization that was particularly good as "East-West" traffic grew in data centers. In addition, the need for improved wireless communications filters in cell phones accelerated the growth of bulk acoustic wave devices.

What about companies that did acquisitions in order to grow and diversify their product mix? Intel is a good example of a company that had an extremely high concentration of revenue in the microprocessor business aimed at PC's and servers (Figure 6). A series of acquisitions in new areas like McAfee for security, Wind River for embedded software, Altera for FPGA's, as well as an organic diversification thrust with the foundry business, added to revenue but not to profit.

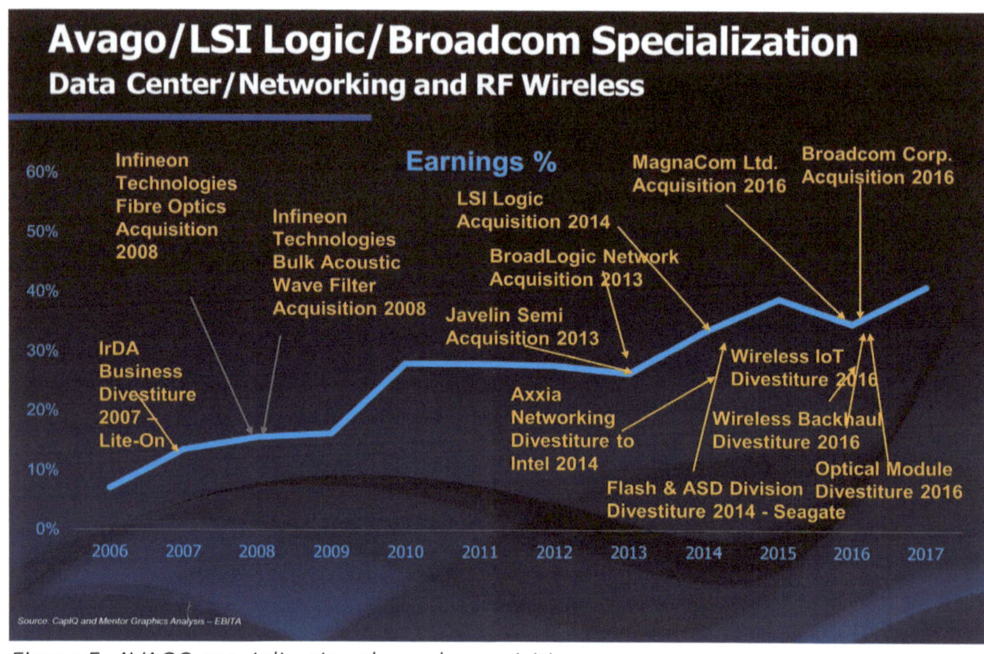

Figure 5. AVAGO specialization through acquisitions.

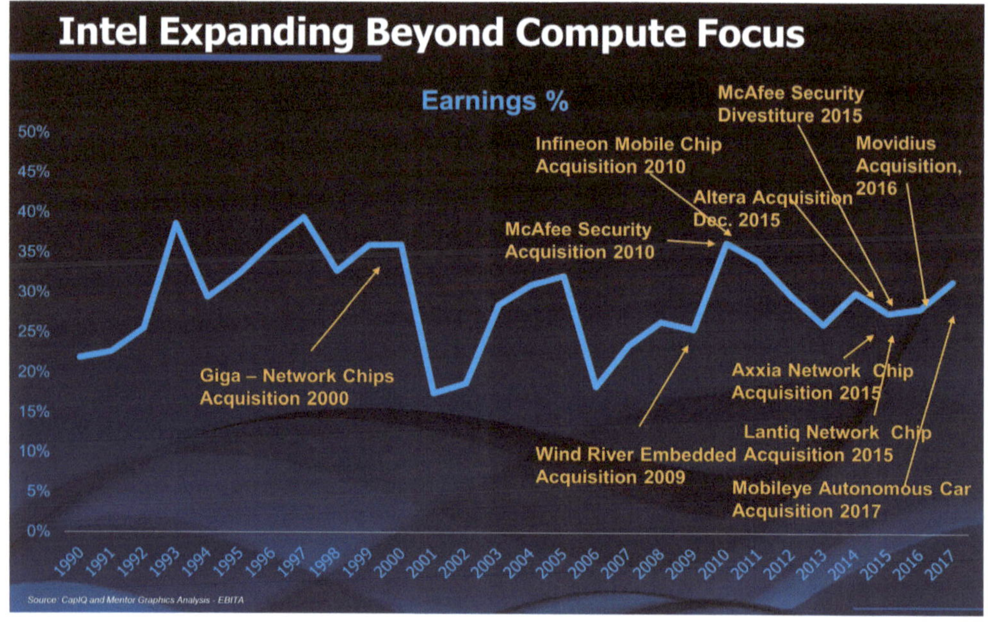

Figure 6. Intel diversification versus profitability.

Finally, one might wonder whether this high correlation of specialization with profitability came as a result of reductions in research and development, especially when one examines cases like AVAGO where substantial cost reductions followed each acquisition. If this did happen, it's not evident for the overall semiconductor industry. The total R&D investment of the semiconductor industry has grown almost every year in history (Figure 7).

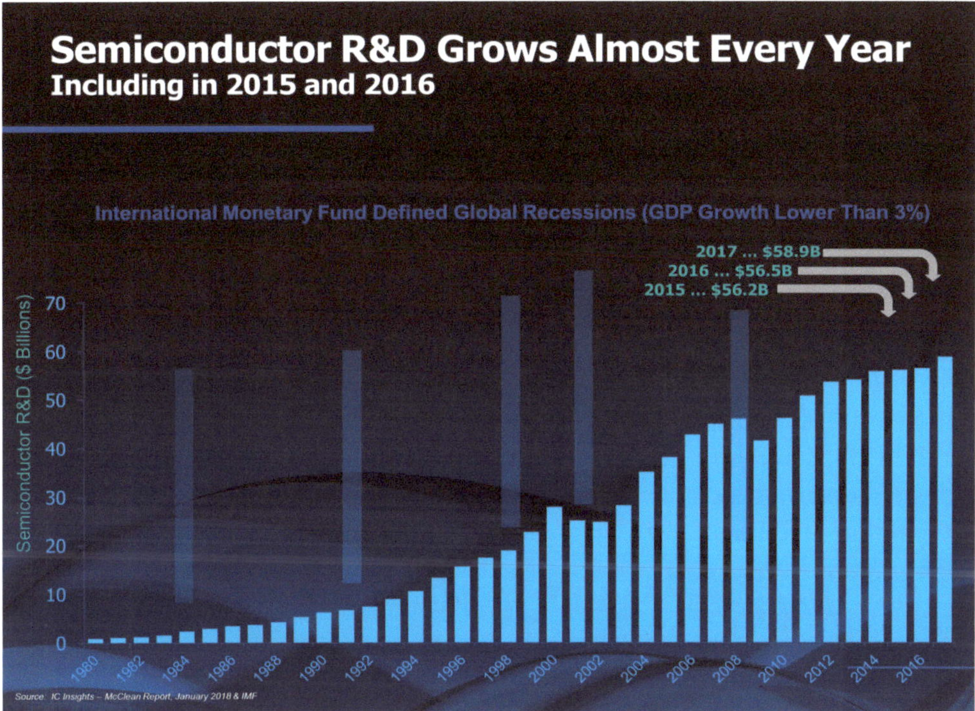

Figure 7. Semiconductor research and development expenditures with recessions shown in gray.

Research and development spending of the semiconductor industry has been relatively constant at 13.8% per year (Figure 8 in Chapter 2). It appears that the managers and investors in semiconductor companies don't believe that their industry is consolidating into a slow growth, mature business. Why would they invest nearly 14% of their revenue each year if they believed that the recent compound average growth rate below 3% was likely to continue?

The semiconductor industry has reinvented itself periodically through history as new applications have evolved. These new applications have created opportunities for new companies to emerge and for the total industry revenue to grow. That's likely to be the case for the foreseeable future.

Chapter 7: Competitive Dynamics in the EDA Industry

Electronic design automation, or EDA, became the term used for computer software and hardware developed to aid in the design and verification of electronics, from integrated circuits to printed circuit boards to the integrated electronics of large systems like planes, trains and cars. As the EDA industry evolved, certain dynamics of change became apparent.

Three large companies have led the EDA industry in each of its eras of growth. Computervision, Calma and Applicon were the three largest engineering workstation companies of the 1970s. They provided special purpose computer workstations for designers to capture the layout of integrated circuits and printed circuit boards and to edit that layout until the designers were satisfied. For all three companies, much of their business came from mechanical CAD (Computer Aided Design) applications but electrical design applications grew as a part of their revenue. The "GDS" standards of today's IC design came from Calma.

In the early 1980s, automation was applied to more than just the physical layout of chips and printed circuit boards. Circuit schematics were captured and simulated on special purpose computers in addition to being transformed into physical layouts of "wires" connecting the circuit elements.

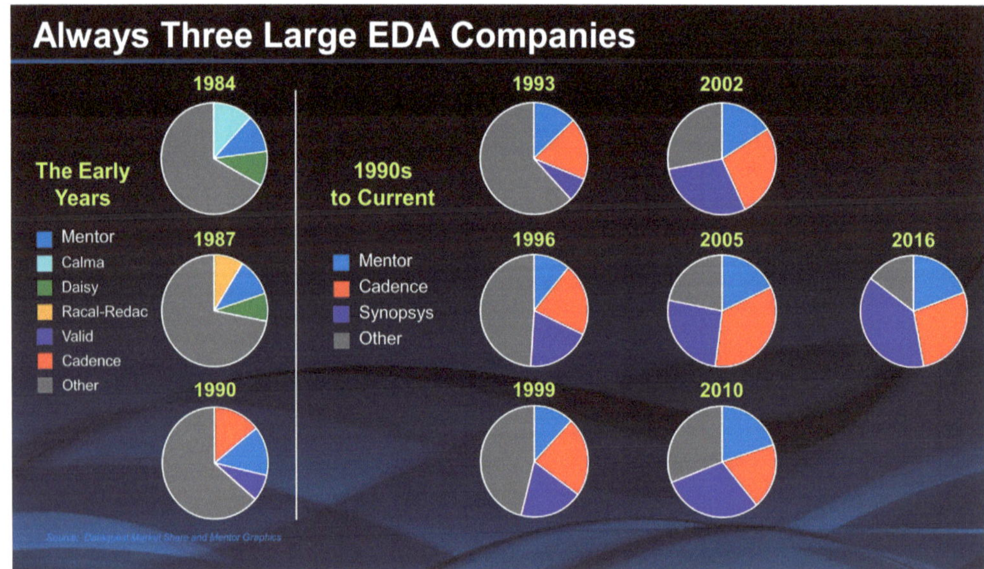

Figure 1. Successive oligopolies in EDA.

Daisy, Mentor and Valid took over the lead during this next decade as Calma and Computervision faded (Figure 1).

Large defense, aerospace and automotive companies selected one of these three companies for standardization across the diverse operations of their own companies. Over time, Valid focused primarily on printed circuit board (PCB) design while Daisy and Mentor did both PCB and integrated circuit (IC) design. Daisy and Valid developed their own computer hardware while Mentor was the first to "OEM" third party hardware, adopting the Apollo workstation and developing software to run on it.

Although this triumvirate had the leading market share through much of the 1980s, the cost and resources required to develop both hardware and software dragged Daisy and Valid down while Mentor survived. Mentor was founded in 1981. Soloman Design Associates (SDA) emerged in 1984 and transformed into Cadence in 1988. Synopsys emerged in 1988.

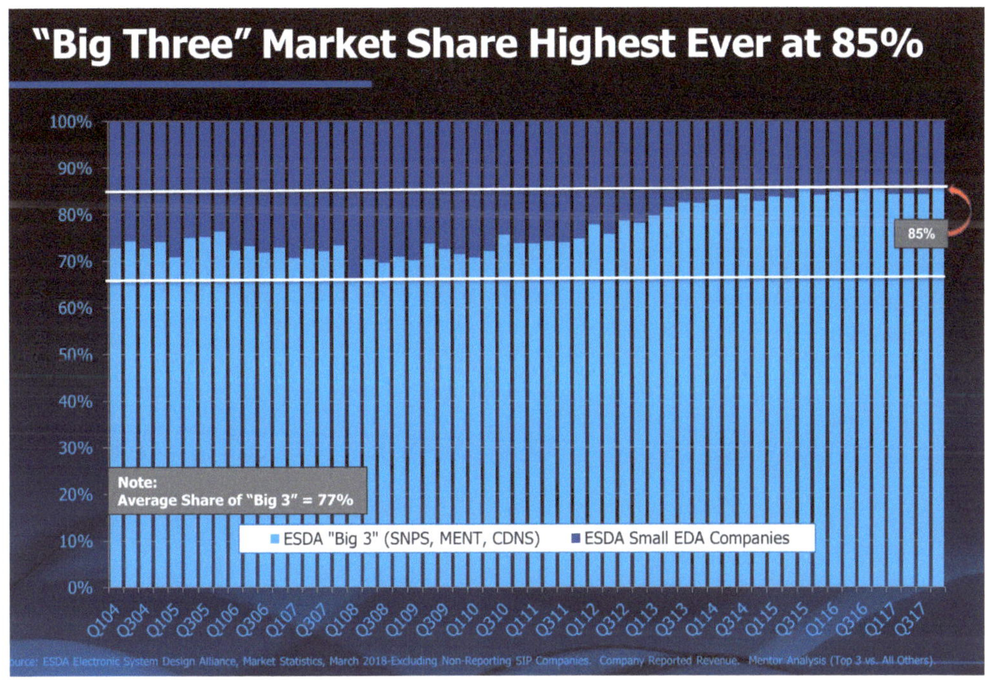

Figure 2. "Big 3" oligopoly with the largest combined market share.

Since the late 1980s, Mentor, Cadence and Synopsys have been an oligopoly with combined market share of 75% plus or minus 5% for most of the 1990s and the next decade. More recently, that percentage has increased to nearly 85%. While Mentor, Cadence and Synopsys had 75%, the other 25% was shared by dozens of smaller companies (Figure 2). Mentor accelerated its market share gains after acquisition by Siemens in 2017 and continued to grow much faster than the market in 2018.

While three companies dominated the EDA business through most of its history, the products making up the revenue of the industry were diverse. GSEDA, one of the leading statistics organizations that tracks the industry, reports on the revenue for 65 different types of products (Figure 3).

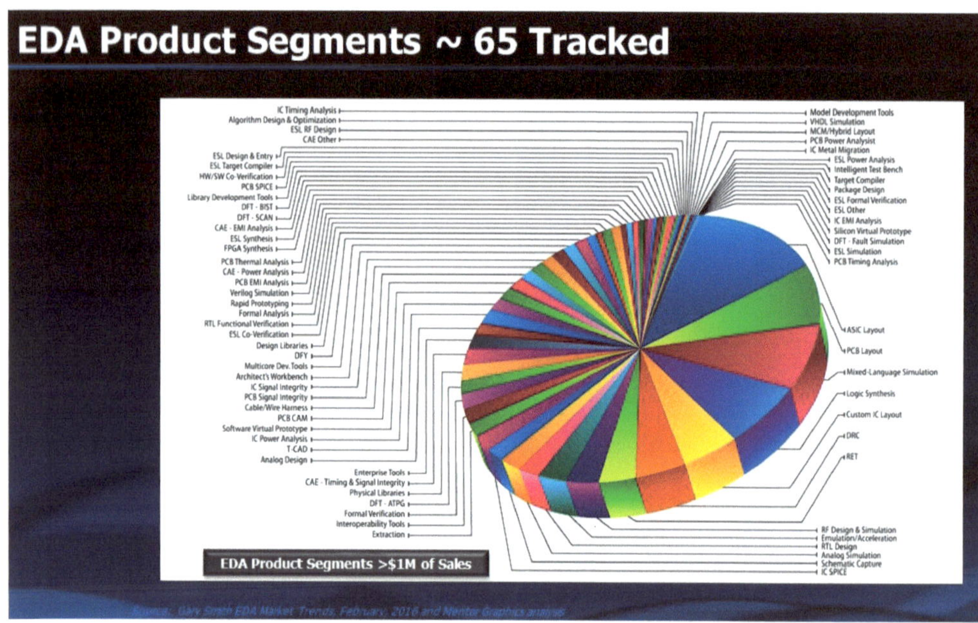

Figure 3. Sixty-five product segments tracked in EDA. Big companies dominate the big segments and little companies dominate the little segments.

Forty of these segments generated $1 million or more of revenue annually. One would think that it would be very difficult for dozens of small companies with very specialized EDA products to survive when three big companies dominate. The big companies, however, dominate the big market segments and the little companies dominate the little market segments. From time to time, little companies are acquired by the big ones.

Overall profitability for the industry remains high because, within any one product category, there is a dominant supplier. Switching costs for a designer to move between EDA suppliers is very high, given the infrastructure of connected utilities and the extensive familiarization required to adopt a specific vendor's software for one of the design specializations.

In the forty largest segments of EDA that generate $1 million or more of revenue per year, the largest supplier in each category has a 71% market share on average. Almost no product segment has a leading supplier with less than 40% market share in that segment (Figure 4).

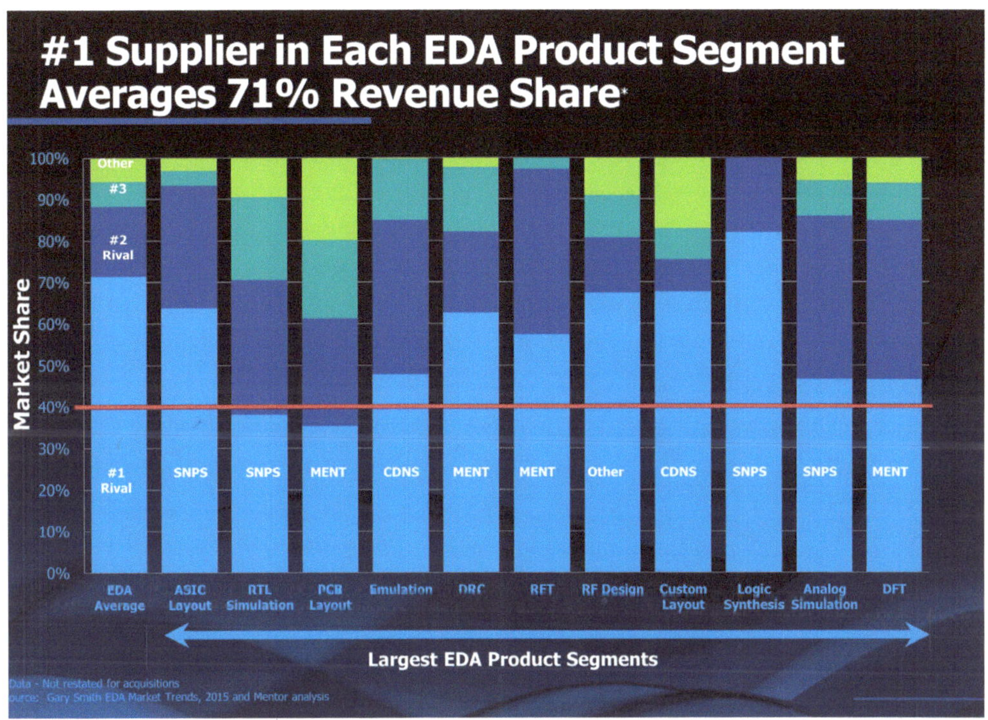

Figure 4. Largest product categories of EDA have a #1 supplier with 70% market share. Minimum market share is about 40%.

As a result, EDA companies with two thirds of the market in any given segment can spend far more on R&D and support in that segment than their competitors.

This gives rise to stability. Engineers are reluctant to change the design software they use because they are familiar with the intricacies of each tool.

Since one EDA supplier usually has a commanding market share in each tool category, that company tends to become the defacto supplier of the tool for that particular application. High switching costs drive stability and profit for the EDA industry and most market share gains come from acquisitions.

Companies that use the tools have the task of integrating design flows using different vendors' tools. Although sometimes difficult when EDA suppliers make it so, this integration is worth the effort to have a "best in class" design flow made up of best-in-class tools. Defacto standards abound and most users find that life is too short to use the tool that few others use. For decades, Synopsys has been the defacto logic synthesis supplier, Cadence for detailed physical layout, Mentor for physical verification and so forth.

Figure 7 of Chapter 2 shows that EDA industry revenue has been 2% of semiconductor industry revenue for over 25 years. Why doesn't it increase as needs and applications grow? Or why doesn't it shrink when R&D cost reduction becomes necessary in the semiconductor industry. First, semiconductor industry R&D has been nearly constant at about 14% for over thirty years. During the 1980s, EDA software costs rose to two points of that 14% by reducing other R&D costs such as labor. Ever since then, EDA budgets have been set so that they averaged one seventh of the R&D expense of semiconductor companies, or 2% of the total revenue.

I'm convinced that salespeople for the EDA industry work with their semiconductor customers to provide them with the software they need each year, even in times of semiconductor recessions, so that the average spending can stay within the budget. As was discussed in Chapter 2, increasing the percent of revenue spent on EDA would require a reduction in some other expense. Rather than do that, the semiconductor industry has unconsciously kept all suppliers on the same learning curve that is parallel to the learning curve for semiconductor revenue. EDA software cost per transistor is decreasing approximately 30% per year, just as semiconductor revenue per transistor does.

What has all this automation done for the semiconductor industry? Figure 13 of Chapter 2 shows the productivity growth per engineer. The number of transistors manufactured each year per electronic engineer has increased five orders of magnitude since 1985. I can't think of another industry that has produced that level of productivity growth.

Chapter 8: Value Through Differentiation in Semiconductor Businesses

Information and figures in this Chapter are covered in greater detail in a You Tube video entitled "Value From Differentiation" presented at the ARM TechCon in 2011 and available at https://www.youtube.com/watch?v=xczYSdz63eU

Gross Profit Margin Percent Provides a Measure of Product Value

The difference between what customers will pay for a physical product and what it costs to make or acquire it is a good measure of differentiation. This difference divided by the revenue is the gross profit margin percent or GPM%. Once a product has an established market share, difficulty of switching to another product usually maintains or enhances the GPM% even if the product differentiation diminishes.

At the time the Apple iPhone was introduced in 2007, it was priced at $749 "unlocked" (i.e. without service provider subsidy) compared to commodity cell phones that sold for $15 at the same time (Figure 1).

Customers Pay a Premium for a Differentiated Product

Figure 1. iPhone differentiated value versus commodity cell phones.

Why would anyone pay fifty times more for a product that did the same thing – made phone calls and sent text messages? Apple proved that there was much more differentiation possible.

Apple's previous attempts with differentiation encountered some difficulties (Figure 2). Before the IBM personal computer was introduced, Apple was able to capture one third of the PC market and maintain a GPM% between 40% and 50%. After the Macintosh introduction, Apple returned to 50% GPM for a while but its market share had dropped to about 20%. Apple then introduced the Mac Classic priced to compete with IBM.

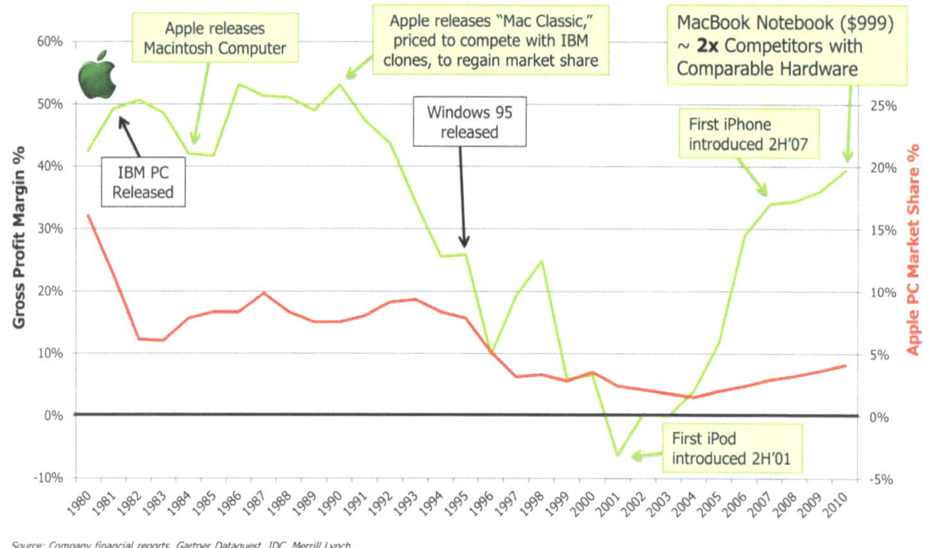

Figure 2. Apple GPM% and market share.

This and the introduction of Windows 95 by Microsoft drove Apple's GPM% down to 10% along with its market share. As Apple's market share subsequently declined to 5% shortly after 2001, GPM% became negative. The iPod and iPhone reversed this corporate trend, bringing the MacBook along with them and resulting in a corporate GPM% of 40% in 2010 and a $999 price for a MacBook that was roughly twice the price of equivalent IBM-compatible PCs. Even today, Apple PCs sell for nearly double the price of the IBM-compatible PC of "equivalent" capability.

Apple's differentiation after 2001 came from more than just the products themselves. Apple built an interdependent ecosystem that made it difficult to switch suppliers if you committed to any part. The infrastructure included iTunes, Apple stores, Genius Bars and MobileMe/Cloud. But the best part of any ecosystem is the part that isn't paid for by the supplier. In Apple's case, that meant the connectors in cars and hotels, the third-party apps on the iPhone and Mac, the peripheral devices, and many more (Figure 3).

Differentiation of Future Products Builds upon Existing Infrastructure and Ecosystem

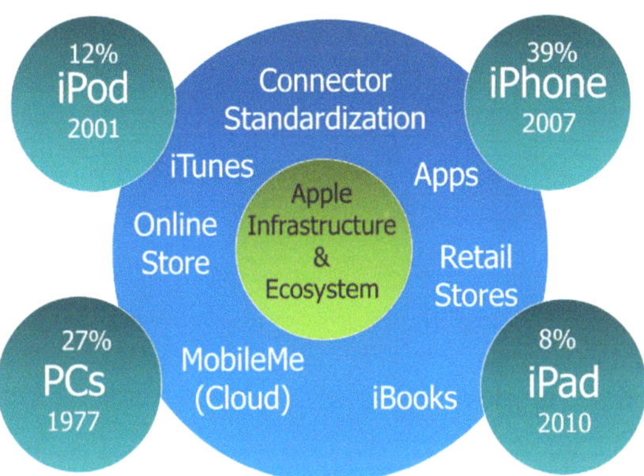

Figure 3. Building an ecosystem to make switching more difficult for existing users.

End product companies are not the only ones who have achieved differentiation that translates into GPM%. Even component suppliers can do this (Figure 4). Intel's 8088 16-bit microprocessor was arguably the worst of its generation. Clearly superior products were introduced (within a year) by Motorola and Zilog.

Combining "first to market advantage" with some legal protections through extension of copyright law to semiconductor photomasks, Intel was able to create a highly differentiated dynasty of products with high switching costs that today dominate the computer server business.

Apple Isn't the Only Company with Sustainable Differentiation

Intel x86 Microprocessors Differentiation

Product	Infrastructure	Ecosystem
Copyrightable instruction set/firmware	Application software/support	Microsoft x86 software base
Patents	Application engineers	Third party training and support
Manufacturing process	Development tools	Third party PC and server manufacturers
Design features		

Figure 4. Intel component differentiation.

Can a Commodity Product Be "De-commoditized"?

Before all the suppliers of "commodity" products become depressed, let's consider the question of "de-commoditization". Don't give up hope. There are lots of examples. Consider Figure 5. Water is generally available for $0.0002 per glass. Yet, in the last few decades, innovative companies have found ways

"Decommoditization" of Water

- Total available market for bottled water is forecasted to be $65B in 2012...nearly 13X the EDA market

- But no algorithms, no engineers...and stable design rules

Figure 5. De-commoditization of water.

to sell water for $0.75 per glass. A similar, but less extreme, scenario could be described for coffee.

I was personally involved in one of the more notable cases of de-commoditization in electronics. In 1977, I was Engineering Manager for Texas Instruments' (TI) Consumer Products Group. TI had aggressively reduced the component count of a basic four-function (add, subtract, multiply, divide) calculator from 480 components in 1971 to a single chip in 1977 (Figure 6).

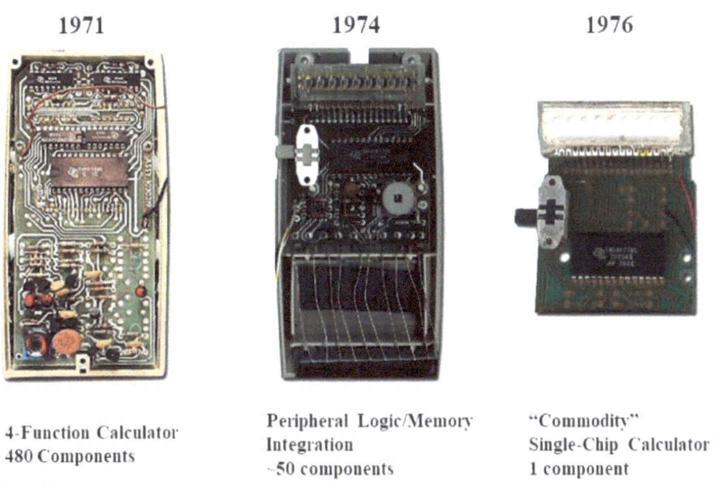

Figure 6. Innovation reduced component count of electronic calculators from 480 to one component in five years.

Pretty impressive, don't you think? TI must have made a lot of money from those innovations? Actually, no, they didn't. The four-function calculators continuously lost money although the scientific calculators made up for much of the loss.

The losses became so painful that TI finally resorted to OEMing (selling calculators manufactured by other companies) as shown in Figure 7.

De-commoditization wasn't easy. But TI had an example to follow from early history with a product called the "Little Professor" which was designed to teach arithmetic. It had a life of Its own. Unlike other calculators, the sales

TI Becomes Uncompetitive When Simply "OEMing" Product from Asia

Source: Datamath Calculator Museum

Figure 7. TI begins "OEMing" calculators made by foreign manufacturers.

continued despite recessions and obsolescence. We learned that parents will pay whatever is required to give their children an advantage in school.

Armed with this history, the TI Consumer Products Group focused on the education market. Over the next two decades, they worked with teachers, school boards, educators, course developers and the entire educational ecosystem to develop a program geared to their graphing calculators.

Figure 8 shows TI calculators that are both differentiated and commoditized. The TI-89, shown on the left, sells for $150. I have personally bought six of these. My two daughters lose them, loan them to friends, break them or require different versions for different classes. TI probably OEMs them for less than $20 each.

For TI, a money losing calculator business has now become one of its most profitable businesses. Figure 9 shows the reported segment GPM% for TI calculators. In 2007, GPM % for TI calculators had reached 65%, at which point TI stopped reporting this segment. As a former TI executive, I have concluded that the reason TI stopped breaking out the reporting of this segment is that it became embarrassingly profitable – more so than the semiconductor component businesses of TI.

Today, TI Calculators Are Both Differentiated and Commoditized

Figure 8. Differentiated and commoditized TI calculators.

Calculator GPM% Dramatically Improves
15 Point Increase over 7 Years

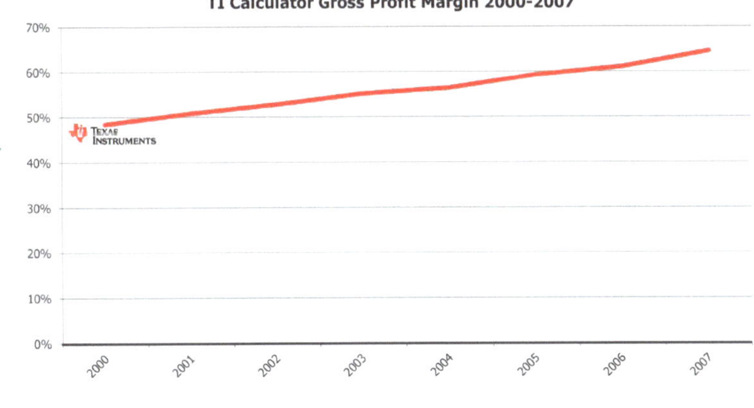

Figure 9. TI's education focus leads to an Increasingly profitable calculator business.

What about semiconductors? Historically, the industry grew with very limited differentiation. Faster, better germanium transistors were quickly matched by competitors because the original AT&T licensing program created a level playing field for semiconductor patents. TI gained a two-year advantage with the silicon transistor but that was soon matched as well. Integrated circuit designs were easily copied and customers had sufficient power to force semiconductor suppliers to have an alternate source of supply for their designs before the customer would "design in" the product.

Over time, semiconductor companies found ways to create differentiation. The most common type is the differentiation that comes with analog components because they can be differentiated by both the design and the manufacturing process. As a result, pure analog companies like Analog Devices and Maxim have demonstrated consistently high GPM%. Most knowledgeable people are surprised to learn that analog components are not the highest consistent GPM% products — field programmable gate arrays, or FPGAs are. Like analog components, FPGAs carry high switching costs once they are designed into a product (Figure 10).

Difficulty of Switching Suppliers Is Greatest for Field Programmable Logic

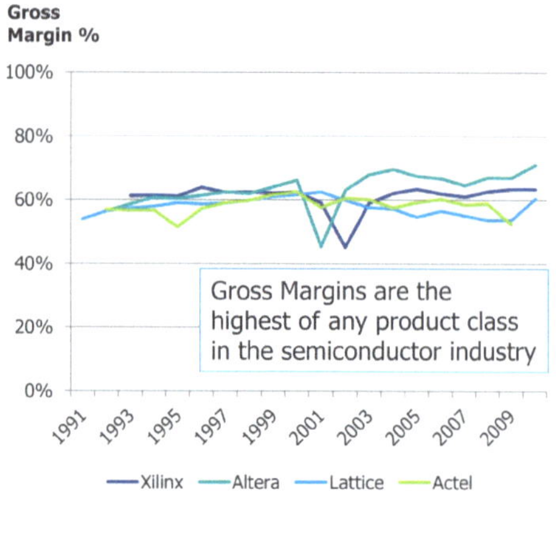

Figure 10. FPGAs have consistently provided the highest GPM% in the semiconductor industry.

FPGAs have an additional barrier to switching suppliers that extends across the entire component product line — differences in the way different manufacturers' FPGAs are programmed creates a barrier for designers to switch suppliers. In addition, libraries of proven reusable blocks of a design are built up in a company over time and are difficult to re-create.

Next highest GPM% among major semiconductor categories, after FPGAs and analog, is the microprocessor. These can be differentiated by their computer architectures and embedded software microcode that can be copyrighted. Patents in the semiconductor industry are difficult to enforce because most large companies are cross-licensed and, even when a patent lawsuit is successful, the precedents for royalties are small, usually less than 5% of product revenue, assuming the licensee doesn't have any patents that can be used as bargaining chips against the licensor.

Intel's de facto standard 808X microprocessors would probably be commodities today if Intel hadn't followed a unique path, i.e. enlisting the forces of the U.S. government to extend copyright legislation to the photomasks for semiconductor components in 1984. The act of copying physical objects that are copyrighted had always been protected before 1982. If you saw a building you liked, you could build one just like it and incur no liability unless you stole the plans from the architect or owner.

Intel's "Semiconductor Chip Protection Act" of 1984 prohibited direct physical copying of chips, even though it had been standard practice in the industry until that time. This limited competing 808X microprocessor suppliers to the Intel-licensed source, Advanced Micro Devices (AMD), and companies that utilized Intel cross-licensed companies like TI for manufacturing. Cyrix was one example of a company that circumvented Intel's barriers by using a licensed manufacturer, TI. In addition, Intel was careful to keep their pricing near the learning curve so that competitors like AMD could be held to a small market share.

Figure 11 shows the GPM% of major semiconductor product categories. There is GPM% differentiation within each category due to other factors such as existing market share. Categories with high switching costs, like FPGAs and microprocessors, tend to have one or two suppliers at the high end of GPM% and others below.

Semiconductor Gross Margin by Category 2009-2010 Average (Weighted Average)

	Company Count Product Revenue >20% within specific SIA category	Gross Margin Average	Gross Margin Median
Foundry	3	39%	24%
Discrete	7	33%	32%
Memory	10	29%	30%
ASSP	28	47%	47%
Analog	12	52%	55%
Micro Component	8	56%	47%
FPGA	3	66%	69%
Total Database	62	47%	45%

e: *Semiconductor Companies (Largest semiconductor companies with published financial metrics available for required reporting periods)*

Figure 11. 2009 – 2010 GPM% for major semiconductor component categories.

Commodity products like memory and discrete devices are more difficult to differentiate partly because their commodity nature is core to the value perceived by customers. Most customers wouldn't design in a DRAM that has no "pin compatible" alternate sources.

Increasing Semiconductor Component Differentiation

What can semiconductor companies do to increase their average GPM%? Figure 12 offers some suggestions. Semiconductor companies have a difficult time differentiating on price, quality of support, sales distribution and many approaches that work in other industries. In general, even manufacturing process differentiation is difficult to sustain for more than one technology generation. Design offers more opportunities to differentiate, especially with "system on a chip", or SoCs, that are sourced by only one supplier. If they incorporate complex algorithms and copyrightable microcode, so much the better. Combinations of process and design, as with analog, RF and power devices, is even better. This benefit accounts for the general movement of companies like TI, NXP and others to an analog-rich portfolio.

Reusable IP blocks are becoming commoditized. There is still an opportunity, however, to develop proprietary IP blocks, as companies like Qualcomm have done, to facilitate superior performance or time to market.

Semiconductor Differentiation in the Future

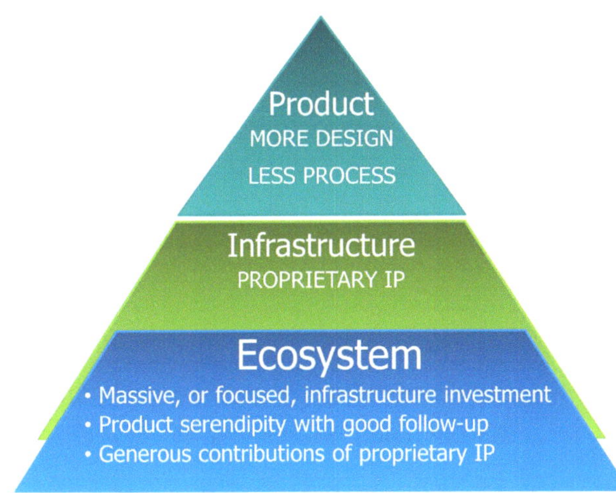

Figure 12. Other ways that semiconductor component companies can enhance GPM%.

Finally, the best differentiation is that created and paid for by customers. Arm is a successful example of a company that built a niche in low power embedded microprocessors for wireless handsets and then expanded to a variety of other applications. Third parties provide a wide variety of assistance, interfaces to other IP, endorsements, etc.

Another example is the effectiveness of open sourcing of a product that increases the value of a proprietary one. Adobe did this with Acrobat. In semiconductors, one of the most impactful moves was from TSMC. Until the late 1980s, silicon foundries kept their "design rules" secret. Customers signed non-disclosure agreements just to find out the information needed to evaluate the viability of a particular foundry for the capabilities the designer needed. TSMC management was frustrated by the share of their business that came from designs that were developed using someone else's design rules, forcing TSMC to tweak their process to match the results provided by the other foundry.

Compass Design Automation, a subsidiary of VLSI Technology, even provided a design library called "Passport". It was popular with designers because, if you used the cell libraries in Passport, multiple foundries had a manufacturing flow that would accommodate the design and produce the same results as the simulated ones. TSMC went one step further by temporarily relaxing secrecy restrictions on their design rules. Compass found it easier to adopt the TSMC design rules for their library, thus solving TSMC's problem. Now all the other foundries had to tweak THEIR processes to match the TSMC results. Effectively, a large share of the foundry customer base became standardized on TSMC's design rules and process.

For the future, the biggest differentiation challenge of the semiconductor industry comes with the Internet of Things. IoT sensors, actuators and controllers are projected to sell in very high volumes at very low prices. Achieving reasonable GPM% is difficult. In the world of IoT, the profit goes to the owners of the information collected from the network of sensors and data collection sites rather than to the providers of the IoT components. As a result, companies like Google, Amazon, automotive OEMs and others are designing their own chips to deploy in information collecting networks.

In these environments, the same company can design and own the IoT sensors and the information collected. Semiconductor companies with IoT component businesses are trying to figure out how to couple their design and manufacture of the components in a joint venture with those who analyze the data. This is a difficult sale so it's likely that we will see continued entry of systems companies into the world of SoC design.

Chapter 9: Specialization Inhibits System Level Optimization

Solving critical customer problems sometimes isn't enough. One of my most interesting experiences came during the development and rollout of a product that was designed to optimize integration of hardware and embedded software. In this case, the product performed exactly as planned but the plan ignored the organizational complexities that come with specialization of skills in different divisions of a large company.

The product, called ASAP (not a great name but that wasn't the reason it failed), analyzed a customer's design at the RTL functional level, along with the embedded software. It determined where the bottlenecks existed for optimum performance or power of the system. We found an ideal customer who was designing a portable consumer product that was dissipating 8.5 mW and wasn't viable with the required size of batteries. Three engineers had worked for a year trying to reduce the power and had modified the design to dissipate only 6.5 mW, still far from the required 4.5 mW.

We analyzed the customer's design and, within a few hours, generated changes that reduced the power to 4.1 mW, well below the 4.5 mW goal of the customer. This was done by identifying bottlenecks and automatically synthesizing hardware to substitute for functions that were inefficiently executing in software on an embedded CPU (Figure 1).

Optimizing Software with Hardware
3 Person-Years of Work Reduced to One Day

SW only (0.11 mW/MHz x 77.0 MHz) =			**8.470 mW**

Co-processor	Power Est	Time Active	HW power Est
FFT	24.0 mW	0.4%	0.096 mW
Fixed pt divider	6.2 mW	0.3%	0.019 mW
Gaussian calc.	5.5 mW	4.7%	0.268 mW
Hamming window	0.2 mW	2.1%	0.004 mW
Total power consumed by Coprocessors			**0.387 mW**

CPU Power Consumption 0.11 mW/MHz × 33.6 MHz	**3.696 mW**

Hardware + CPU =	**4.083 mW**

Does *not* include potential power saving from voltage scaling

Figure 1. Automatic analysis and synthesis to achieve reduced power.

The customer was ecstatic about the result and we expected a major sale. When we didn't receive an order, we investigated. The problem, it turned out, was an internal disagreement. While the customer engineers agreed that they badly needed our product, they couldn't agree which group, hardware or software design, would be responsible for using it.

The hardware engineers were adamant that "no software engineer is going to generate hardware in my chip design" and the software engineers were adamant that "no hardware engineer is going to change a single line of my software". Amazingly, the disagreement was so strong that they decided not to adopt our product and, instead, to kill the development of their own very promising product.

You might think that this was an extreme example but I'm increasingly convinced that it wasn't. We experienced the same thing every time we developed products that crossed domains of expertise, from analog to digital, from mechanical to electrical, from software to hardware, from design to manufacturing, etc. Software tools that appealed to one domain were not accepted by the other domain. (Figures 2 and 3).

Figure 2. Each discipline has its own culture, language, perspective, and metrics. Differences in the way specialized groups do their work make it difficult to provide tools that cross domain boundaries.

Will Organizations Ever Find a Way to Optimize Across Domains of Expertise?

Figure 3. Differing standards, metrics of performance, modes of communication and other differences prevent system level optimization.

This is a phenomenon that appears repeatedly, especially in large organizations. There are, however, ways that system level optimization can be achieved. Some of these are listed in Figure 4. One of the most apparent examples of the evolution from problematic partitions to a successful organization structure was the change in the customer/supplier relationship that evolved with the advent of silicon foundries in the semiconductor industry over the last thirty years. When most semiconductor companies were vertically integrated, the tradeoffs of every new process technology led to major feuds between the design and the manufacturing engineers. I know this because I had to referee the arguments many times.

Ways to Overcome Organizational Partitions

- Change the customer/supplier relationship
- Create a start up
- Assimilate the task of one group into another
- Move to a higher level of abstraction
- Form an abstraction layer
- Multi-physics simulation/modeling
- Multi-disciplinary data management

Figure 4. Ways to overcome the barriers of organizational partitions.

A new generation of product required the most aggressive feature sizes possible while manufacturing yield and throughput favored the least aggressive. A compromise had to be made and it usually was influenced more by politics than by engineering. With the emergence of silicon foundries, the problem went away. Now there were suppliers whose success depended upon providing the most aggressive design rules possible in a cost-effective manufacturing environment. No more politics. Just an insightful analysis of the manufacturing and design tradeoffs.

Another approach to solving the specialization problem is to form a startup company (Figures 4 and 5). In a typical early stage startup, partitions of specialization have not yet formed, so the hardware engineer also frequently writes some of the embedded software, or is at least heavily involved in both.

Eliminating Functional Barriers

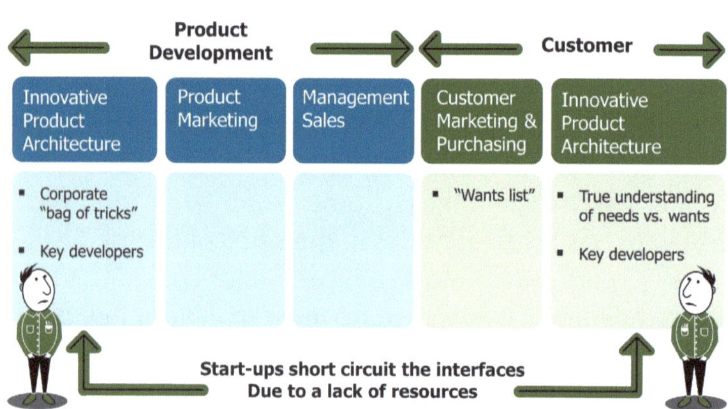

Figure 5. Removing the interfaces between customer and supplier.

In addition, startups typically have a key technical expert who will be respected by the potential customers' most valuable development engineers. The two of them can get together and exchange ideas for the ideal product because the startup engineer is not constrained by finding the solution in only one domain, e.g. in software or hardware. Assimilating the task of one group into another also removes artificial partitions. This assumes that the new group truly integrates the responsibilities. It could mean making a group leader at a low level responsible for an integrated solution that involves both hardware and software, for example.

Another approach is to move the design to a higher level of abstraction so that tradeoffs can be made among differing specialties, e.g. hardware/software, mechanical/electrical, etc. If the abstraction level is high enough, then everyone can speak the same language (Figure 6).

Figure 6. Abstraction temporarily solves optimization challenges.

This works for a while but very quickly, the addition of detail into the design causes a split among specialties and the optimization effort is reduced or ceases entirely. In some industries, a new abstraction layer can be created at a lower level to overcome this problem. SysML is one example. AUTOSAR,for the automotive industry, is another.

Another solution is to conduct multi-physics simulation of designs to see the impact of tradeoffs in different domains (Figure 7). Even with this type of simulation data, it's frequently difficult to determine which design domain should make changes to improve system level performance. As a minimum, however, it provides data for a rational discussion and takes some of the emotion out of the decisions.

While these approaches offer potential, one must wonder whether there are any solutions that are universally applicable? One overarching approach comes from the way a company handles its data management. For years, company managements hoped for the universal workstation that could be used by the many different disciplines—mechanical, electrical, software, etc. That is not likely to happen. Engineers need their own ways of working with design and manufacturing data and they are not going to change, nor would it be advisable to do so. Efficiency in one domain requires different tools and methods of analyzing data that may not be efficient in another domain.

System Modeling with Multiple Variables

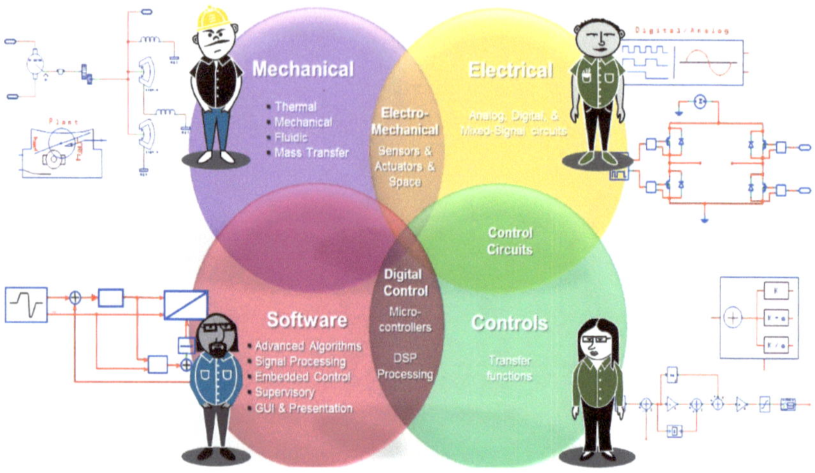

Figure 7. Multi-physics simulation.

Despite this need for separation of development functions, engineers still need information from other domains to do their jobs. A systems company needs a centralized database from which groups in different areas of the company can download and upload data for their own work and to access information from other domains.

A good example is the engineer who is developing wiring for a plane or car. Electrical design of the wiring harness requires detailed electrical simulation, analysis of potential sneak paths and optimization of "take up" alternatives of options in the vehicle so that the basic wiring cost is minimized while the wiring harness can be customized for a multiplicity of option combinations in a vehicle. At the same time, the wiring approach will change to meet the three-dimensional characteristics of the mechanical design of the vehicle.

How does the electrical designer obtain the data needed to determine if a wire bundle will fit through a hole in the frame of the car? Or how does the designer know the wire lengths in three dimensions? Does the designer import the mechanical database? Impractical and probably impossible. An extract of estimated wire tracks and lengths must be exported to the mechanical design environment and then simulated with mechanical models and tools. Similarly, subsets of system design data must be extracted from one design discipline to another throughout the design process evolution.

Over many years, I have had the opportunity to work with teams to develop and modify products to make them usable by developers in different domains of expertise. Some of the lessons learned from this experience are illustrated here.

Lessons Learned: Provide Unique Data Structures and Data Bases for Each Discipline

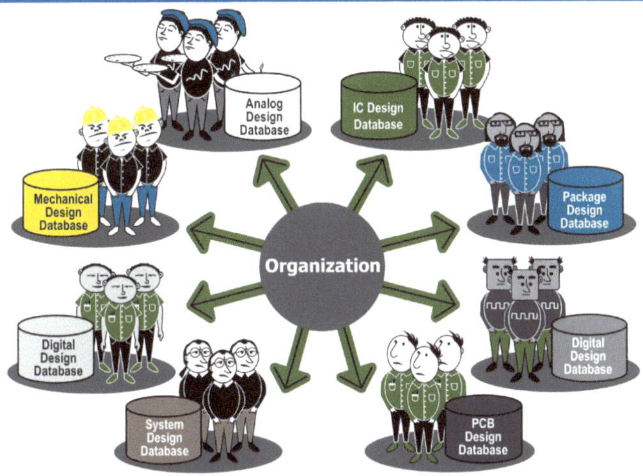

Figure 8. In an enterprise data base, unique data structures are needed for each type of discipline.

First, it's important to provide unique data structures and datasets for each discipline. Mentor's experience with Version 8.0 of our software drove this one home. Forcing all the users to format their data in a fixed set of predetermined formats creates an inflexible system that doesn't benefit anyone but the database vendor. The database needs to be open and flexible.

Beyond Mentor's own disastrous experience with the fixed data formats of the Falcon 8.0 database, we were later forced to support our Capital electrical architecture software on a Catia set of formats that suffered the same problem as Falcon. Performance would have been hopelessly compromised, changes to database structures would require a major regeneration and verification of the database software and our product would have been vulnerable to knockoff by the database owner. Instead, we created a digital flow for our data outside the Catia database.

Figure 9. Don't burden one discipline with another discipline's detailed information.

This approach requires working with data base vendors who favor openness. This has always been a fundamental priority for Siemens Teamcenter and federated data base approaches of other companies but not necessarily for all database providers. Openness was a key compatibility philosophy for the merger of Siemens with Mentor Graphics that made the union successful.

As mentioned earlier, there are still many people who believe that disparate design disciplines in a company should all use the same workstations, the same user interface, the same data structures, etc. This philosophy is driven by the idea that it is good to have a single design and verification environment that transcends the differences in the enterprise. Engineers can then move from group to group with minimal retraining and design information is more

Lessons Learned: Enable Selective Access to the Required Data; Transform Data Formats Quickly and Easily

Figure 10. Enable selective access to the required data; facilitate rapid translation of data formats.

easily shared. Despite support for this concept among the managements of many companies, it rarely, if ever, happens.

Burdening an electrical designer with the overhead of the mechanical, manufacturing, thermal, etc. detailed design information doesn't seem to work. The trick is to be able to access the pieces of data from another domain that are needed to do your job in your domain. Even better is an architecture that lets you export abstractions of your design to another domain to perform tasks not well suited to the domain of your expertise. This is how electrical wiring is done when the electrical designer needs to make sure his design meets the constraints of the mechanical embodiment of the product (Figure 10).

Flexibility and openness of the enterprise data base is the most important criterion (Figure 11). If addition of a new data format requires a major revision of the entire data base system, it's impractical to wait. Typically, other things are impacted when a major revision of this type is attempted so the data base structure must be designed for flexibility to change some formats without having to reverify the entire database system.

Lessons Learned: Change One Area of Development Environment Without Re-Verifying the Entire Enterprise Environment

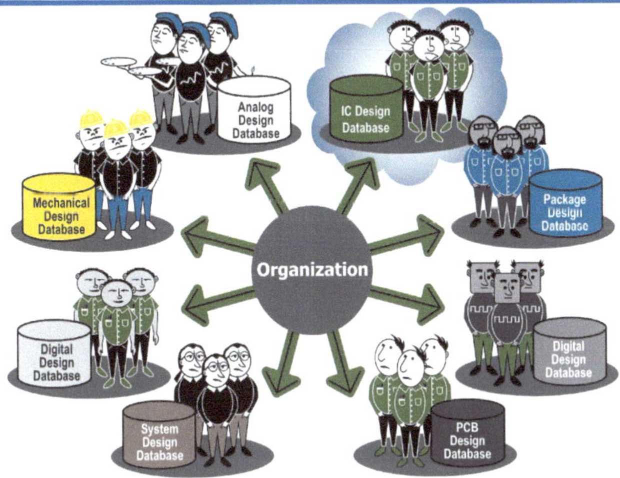

Figure 11. Make sure the enterprise data management has the flexibility to add or change data formats selectively without re-verification of the entire data base management system.

Finally, the more a design environment feels familiar, the more likely the development engineers will create good products (Figure 12).

Lessons Learned: Tailor the Software to Each Discipline

Figure 12. Developers have enough to worry about without adapting to changes in their design environment and support.

Although the "lessons learned" provide guidance for how data bases and design environments should be structured, few large corporations have been able to implement the level of interoperability between disciplines that they would like. Figure 13 is still a hope rather than a reality.

Even if the commercial databases and design software provide the capability for data to be accessed and analyzed from functional domain to functional domain, system optimization would still require that compromises be made in one domain to achieve the optimum result at the system level. Perhaps this is why systems companies who find ways to overcome this challenge have traditionally achieved higher operating margins than component companies.

It's likely that success will evolve application by application. The case of electrical wiring of cars and planes reached such a critical level that integrated solutions evolved among the electrical, mechanical and manufacturing domains. Other applications are reaching a critical point where system optimization can only be achieved in an environment where multi-domain tradeoffs can be made.

Recognizing the Differences in the Needs of Different Design Disciplines is the Solution to World-Class, Enterprise Design

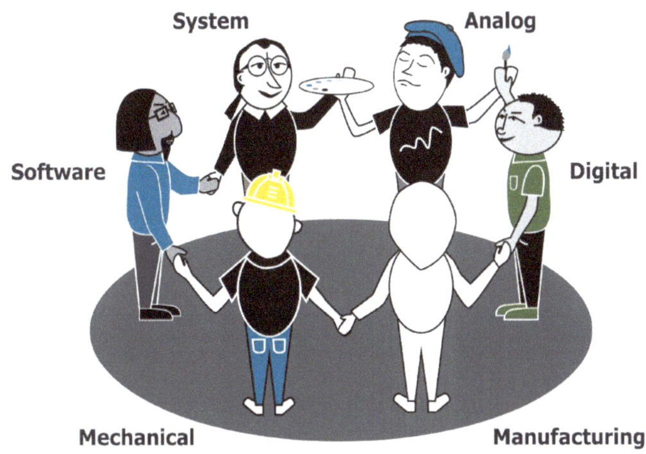

Figure 13. Specialization in large enterprises can be a strength, rather than a burden. Development environments that maintain the needed specialization by discipline while affording access to data in other domains leads to the most productive enterprise.

Making these tradeoffs at the highest possible level of abstraction is most likely to produce an optimum result and is also most likely to facilitate compatible development in the diverse functional domains of the corporation.

Chapter 10: Design Automation for Systems

Electronic design automation has evolved to an extent that the complex chips with tens of billions of transistors frequently produce first pass functional prototypes from the manufacturer. What makes this so incredible is that such a small portion of the possible states of electronic operation are actually tested in the simulation of the chip.

Figure 1 takes the example of a very simple electronic function, a 32-bit comparator, that compares two thirty-two bit numbers and determines whether one of them is equal to, less than or greater than the other.

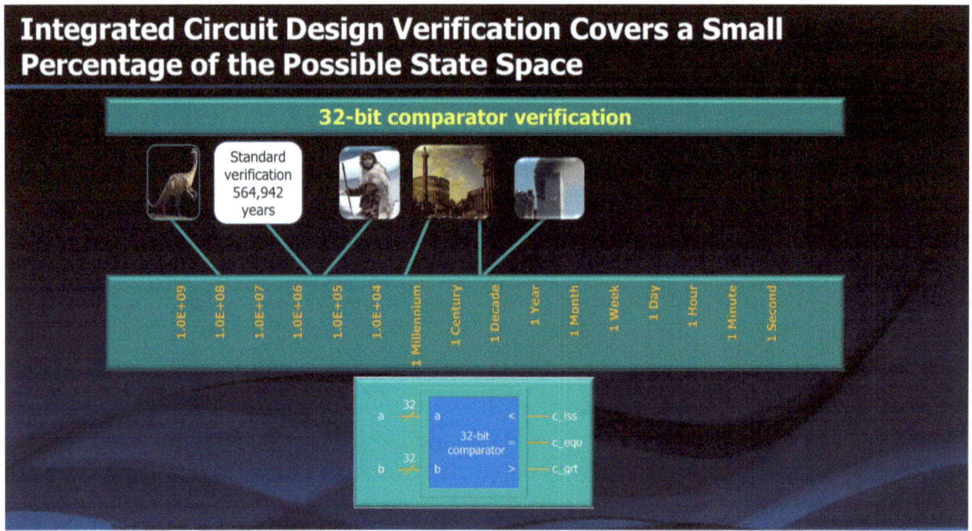

Figure 1. Simple comparison of two 32-bit numbers would require 565,000 years with a state-of-the-art computer if each possible pair of numbers had to be compared.

One might naively assume that this requires 2^32 comparisons of the two numbers. It doesn't. If it did, then a caveman who was given one of today's state of the art computer servers 565,000 years ago would just have completed the calculation. EDA history is made up of innovations that preempt the need to check every possible state of an electronic circuit, or 100% of the state space as design practitioners would say.

The question then arises, "if we can reliably simulate the behavior of chips with billions of transistors, can we extend the technology to more complex systems like cars, planes and trains?" Or, if we can do this for the electronic behavior of a chip, could we extend it to the mechanical, thermal, aerodynamic or other simulated behavior of a complex system? Inverse reasoning suggests that the answer is "yes". The reason is that the electronics of systems like cars and planes are becoming so complex that, if we can't automate the design and simulation, there is no other known solution. Humans certainly can't analyze the complexity of such a system (Figure 2).

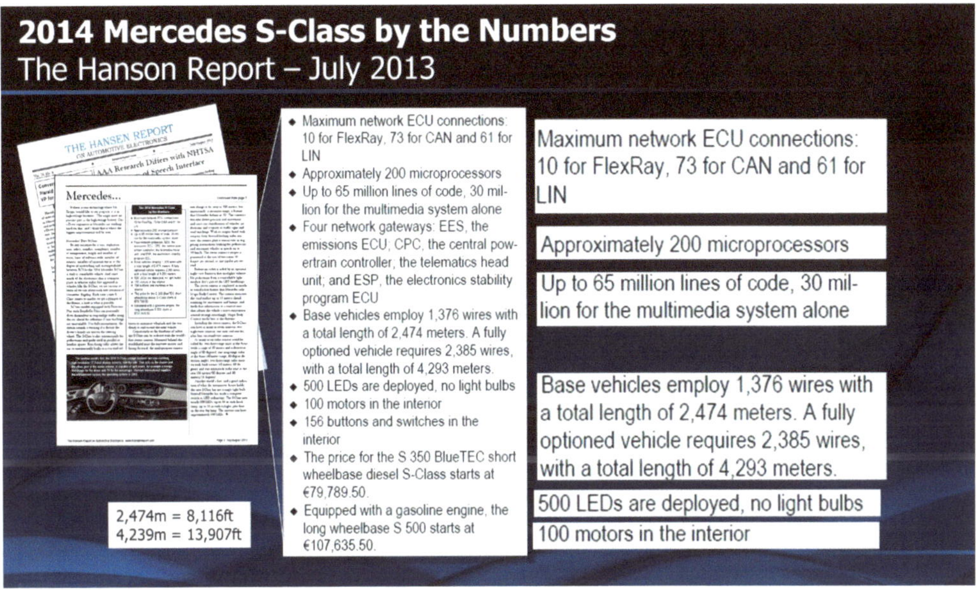

Figure 2. Electronic and wiring complexity of a 2014 S-Class Mercedes.

It has taken sixty years to evolve the software to accurately simulate the electrical behavior of chips. How long will it be before we can do the same for an entire car or plane? And how will cars and planes be designed in the meantime?

For the automotive and aerospace industries, mechanical design simulation and verification evolved long before electronic simulation. Dozens of mechanical computer automated design, or CAD, companies emerged in the last thirty years. Today simulators that model most of the mechanical design, as well as the manufacturing processes to produce them, are available from companies like Siemens, Dassault and Parametric Technologies. These simulators also analyze aerodynamics and thermal effects.

It's just in the last three decades that the electronics in cars and planes have increased in complexity to such a level that humans can no longer manage the data required to create an optimized, cost effective design without errors (not to mention protections against hacking).

It's easy to assume that the design of a car you buy has been verified by driving prototype cars for thousands of miles in all types of weather conditions. It probably has. Before a manufacturer can build that prototype, extensive verification must be performed. How is that done? It all comes down to a methodology called "abstraction". Requirements for the design of a vehicle are described at a high level and then refined to provide greater detail. Each level of abstraction of the data is analyzed on a computer or with a physical prototype of a subsystem.

The same is true of integrated circuits. Figure 3 shows the various abstractions used to describe, simulate and verify the performance of a chip.

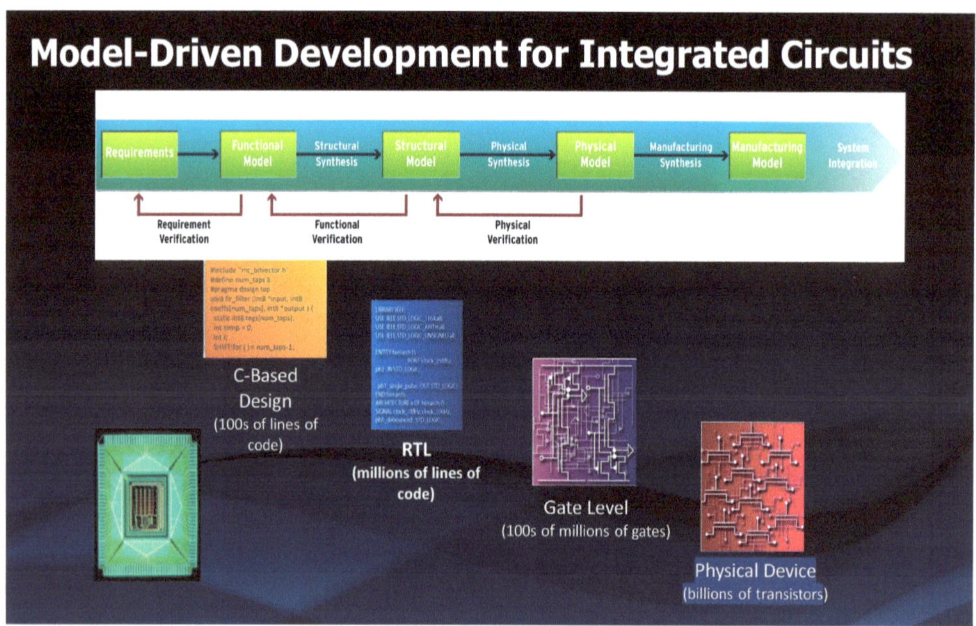

Figure 3. Four "abstractions" used in the design of integrated circuits.

Although relatively new, ICs are increasingly being described in a high level language like C++. This description is relatively compact so simulations of the entire chip, or the critical performance portions of it, can be run quickly. That description is automatically "synthesized" into the next level of abstraction called "RTL" or register transfer logic that is described by a language such as Verilog, VHDL or System Verilog. This level of abstraction is much more detailed, describing logical operations of the chip.

Simulations of the full chip typically take up to twenty-four hours, so the building blocks of the chip are rigorously simulated before integrating them step by step until the whole chip can be simulated. Once the designer is satisfied with the RTL simulation, the database is synthesized into a description of the actual logic gates creating what is called a "netlist". The design is synthesized into a description of the physical layout of the transistors on the silicon and then transformed into a language (GDS2) that the photomask generator can understand and can convert into the actual photographic negative that is used to manufacture the chip.

System design has evolved a similar design approach but systems engineers refer to it as the "V Diagram" (Figure 4).

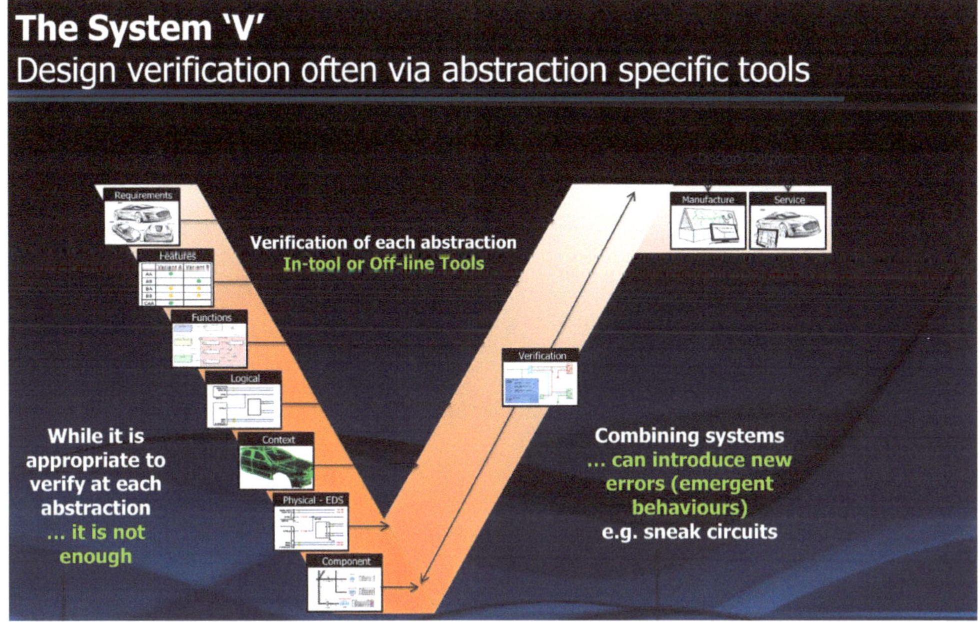

Figure 4. System "V" diagram showing the path from high level abstraction to greater detail followed by integration and verification at each level of abstraction.

A difference between the "V" approach and that used by IC designers is that the system designer is likely to build a physical prototype of each subsystem once the design is refined to the level of a physical description.

That prototype can then be tested by inserting it into a laboratory mockup of the entire vehicle using what is referred to as "hardware in the loop" testing. Integration testing can also be performed with hardware in the loop but increasingly those subsystems are tested in a "virtual" environment where the parts of the vehicle that are connected to it provide inputs and react to its outputs in a simulated virtual environment.

This whole methodology is being disrupted because of growing complexity. Once we begin to develop truly autonomous vehicles, the approach will become totally inadequate because the number of tests that must be performed exceed the capability of physical testing (Figure 5). To test an autonomous drive vehicle would require more than eight billion miles of driving, according to Akio Toyoda, CEO of Toyota.

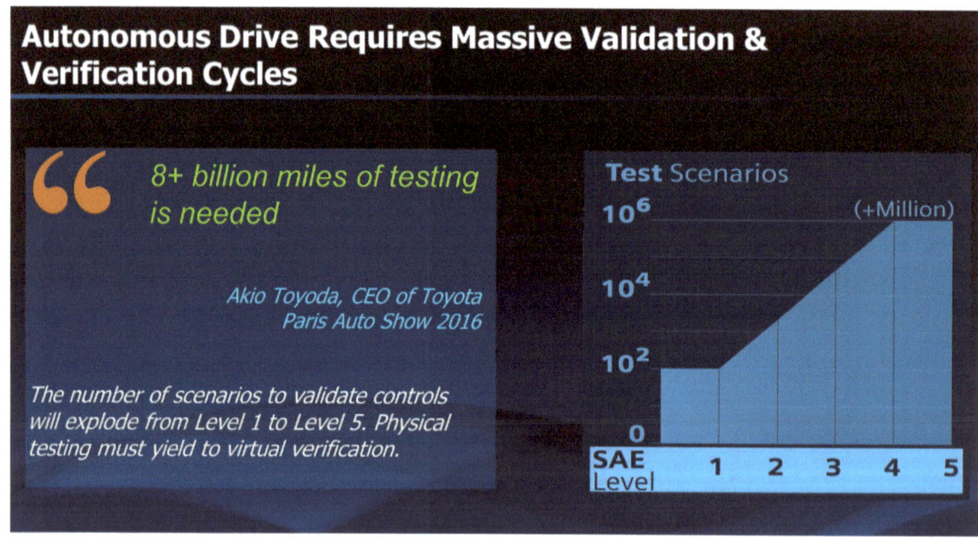

Figure 5. More than 8 billion miles of driving would be required to physically test an autonomous vehicle. Instead "virtual" verification must be adopted.

A manufacturer would have to send out a fleet of 300 cars, driving at 60 mph for fifty years. Not very practical for introducing a new model each year.

Another reason that automotive and aerospace design must become virtual is that optimization has become too complex. Consider the wiring alone. With more than 1.5 miles of wiring in a car, forty miles in a small business jet and over one hundred in a commercial aircraft, there is a critical need to analyze tradeoffs among variables like weight, cost, performance, signal integrity, etc. Finding an optimum combination is far beyond the ability of the human brain.

The same can be said for optimizations of the electrical subsystems, called electronic control units or ECUs in a car, or line replaceable units or LRUs, in an airplane. These ECUs contain multiple chips and embedded software to handle processing such as control of brakes, transmission or engine ignition. They are complex enough to require simulation to assure that the inputs and outputs perform as specified. The additional opportunities for problems arise when the ECUs are tested in a system environment. Even if an automotive OEM were lucky enough to produce a functioning car without a virtual simulation, debug of future problems would be difficult or impossible without a simulation.

Modern cars contain up to one hundred million lines of software code. It's safe to assume that this code will contain bugs. The challenge for the automotive OEM is to find a way to react quickly and update the software in every similar vehicle on the road when a bug is discovered. Otherwise, the OEM could be liable for all accidents that occur once the bug is known. Tesla has developed an infrastructure to make this possible.

The other challenge is to design the car in such a way that mission critical systems can be isolated. Many of the most publicized hacks of vehicles have come from intrusion of the vehicle through the infotainment system that is tied to the CAN bus, giving access to more critical systems like the brakes, transmission and engine.

How long will it take until automotive OEMs design the entire vehicle, as well as the assembly line for building it, in a totally virtual environment on a computer? The industry is farther along than you might think. Most of the mechanical design and manufacturing operations are already done that way. The remaining challenges include much of the electronics. That's why Siemens, who provided software for all aspects of mechanical, aerodynamic, thermal and manufacturing simulation, decided to acquire an EDA company, Mentor Graphics.

System simulation of the electronics, as well as testing and optimization that involves "cross domain" testing among electrical and mechanical systems, remains very challenging. Wiring architectural tradeoffs and automatic generation of the design of the wire harness is essentially automated today. Automation of design and verification of other vehicle electronics will require development of abstractions that can be used to analyze multiple ECUs operating in concert with one another as embedded software is executed in the vehicle. The abstractions must be at a high enough level that they can be simulated at something like 100X the real time execution but be detailed enough that an engineer can analyze the inner workings of an ECU to find a design bug or test an optimization alternative.

How long will this take? Not that long. It has to happen over the next decade or two or we won't be able to design the next generation of cars and planes.

Chapter 11: International Semiconductor Competition

Semiconductor industry evolution was largely a U.S. phenomenon. While there were important contributions made by persons all over the world, the basic technology grew from the invention of the transistor at Bell Labs which was licensed broadly in the U.S. That created a level playing field for all who wanted to become producers. The industry then evolved in a free market environment in the U.S. largely without regulation or patent disputes.

By 1965, three companies, TI, Motorola, and Fairchild, had a combined market share greater than two thirds of worldwide sales of semiconductors (Figure 7, Chapter Five). Barriers to the internationalization of the industry emerged through import tariffs in Europe and restrictions on setting up subsidiaries in Japan but even these limitations gradually disappeared.

After 1965, the semiconductor industry began a near continuous deconsolidation as new companies entered the market. The market share of the largest competitor remained about the same for the next fifty-five years at 12 to 15% (Figures 7 and 8 in Chapter Five). Top ten semiconductor companies during the 1950s and 1960s did not include any non-U.S. companies (Figure 10, Chapter Five).

Japan Becomes a Significant Competitor

In the late 1970s and early 1980s, the industry became international with the entry of NEC, Toshiba, Hitachi and Matsushita into the list of the top 10 largest semiconductor companies. Philips, a European semiconductor company, also entered the top ten for the first time. The Japanese phenomenon was driven largely by the superior manufacturing process control applied to dynamic RAMs, or DRAMs. This was the first wave of "trade wars".

The Japanese Ministry of Industry and Trade, MITI, coordinated actions among Japanese companies that contributed to a cooling of tensions between SIA (U.S. Semiconductor Industry Association) and EIAJ (Electronics Industry Association of Japan). Japanese companies were assigned quotas for purchases of semiconductors from U.S. companies, relieving some of the pressure. The real end to the issue was predictable but not so obvious to many of us.

At a dinner I had with Saturo Ito, CEO of Hitachi Semiconductor, in the late 1980s, he explained to me that the U.S. shouldn't worry about Japan taking over most of the manufacturing of semiconductors in the world, as I had feared. Ito told me, "Japanese are optimizers, not inventors. When standards are stable, Japan will do well. When standards are evolving, not so well." He was right. As the personal computer and cell phone industries grew, the U.S. recaptured substantial innovation momentum.

Enter Korea

Less predictable than the Japanese success in DRAM manufacturing was the entry of Korea. When Samsung announced its intent to design and produce the 64K DRAM and sent their managers on a 64-mile hike as a symbolic start, we didn't pay much attention. How could they catch up in an industry that was so mature?

Their success came from the determination of Koreans when they decide upon a specific goal. The path was not easy. As late as 2008, the combined market share of the three largest DRAM producers in the world, Samsung, SK Hynix and Micron, was only modestly greater than 50% of the entire worldwide market. Today, their combined market share is greater than 95% (Figure 6 of Chapter Five).

Taiwan

As with Japan, Korea and China, Asian governments understood very early the importance of domestic semiconductor capability. Evolution of worldwide leadership in the silicon foundry business by Taiwan is truly remarkable. The Taiwanese semiconductor industry grew from a technology transfer program funded by the RoC. A team of Taiwanese engineers went to RCA in 1976 for the transfer and then returned to Taiwan to work for ITRI, a government supported research institute that was founded in 1973.[1]

These individuals were very talented. They started many semiconductor companies beginning a decade later including UMC, TSMC, Mediatek and many more. Taiwan wisely chose Morris Chang, former head of TI's semiconductor business, in the 1980s to manage ITRI and ERSO, the two principal research entities involved in electronics. When TSMC and UMC were formed, Morris filled the role of Chairman of both.

Securing funding from Philips, he was able to take advantage of an unlikely transition of the semiconductor industry. Those of us in the industry at the time wondered how a company whose business consisted only of wafer fabrication could ever survive. Those of us working for integrated device manufacturers like TI viewed wafer fabs as our biggest problem. Whenever a recession came along, the fabs became under-loaded and were astoundingly unprofitable. The depreciation cost continued but the revenue did not. If someone else was willing to take over the wafer fabs, then the semiconductor business volatility would be reduced, making it a more attractive, stable industry. The U.S. had already experienced the pain of wafer fab ownership when Japanese competitors in the 1985 recession kept running wafers at a loss, partly because they couldn't lay off their people.

In the U.S. we shut down some of our wafer fabs to stop the financial bleeding. Step by step we created an opportunity for Japan to gain market share when the recovery inevitably came. Now TSMC and UMC in Taiwan were taking over the one aspect of the industry that caused us the greatest pain.

TSMC recognized the value of being a dedicated foundry with no products of its own to compete with its customers, unlike its competitors such as UMC, Seiko Epson, NEC and more. By becoming a pure play foundry and popularizing its design rules, TSMC gained momentum that was hard to match. Today, only Samsung comes close to providing real competition at the leading edge of semiconductor technology.

This leaves the U.S. in the interesting position that, other than Intel plus some limited capacity in Texas and New York, there is no significant domestic manufacturing capability at the advanced nodes of semiconductor technology. Anything that disrupts free trade between Taiwan and the U.S. could severely disrupt much of the U.S. manufacturing and defense industries.

China

Despite China's rise as the world's largest assembler of consumer electronic equipment, the Chinese semiconductor industry has evolved slowly. The largest Chinese semiconductor foundry, SMIC, is still two technology nodes behind TSMC in manufacturing capability as of 2019. The Chinese government is dedicated to changing this situation. Figure 1 shows the semiconductor investment announced by the Chinese government in 2014 and 2018.

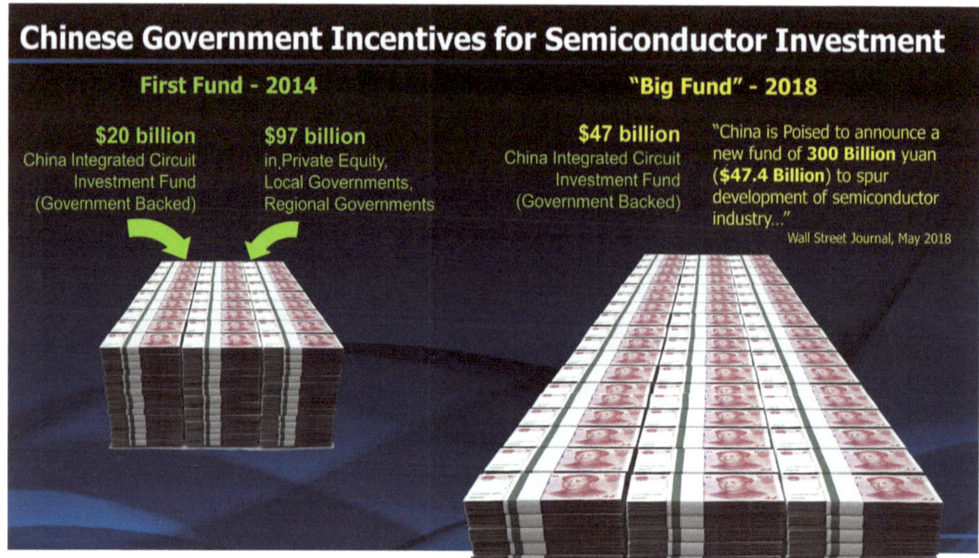

Figure 1. Chinese government stimulus for semiconductor investment has been matched by a much greater investment by private equity companies.

These numbers dwarf anything that the U.S. government is likely to do. The entire annual revenue of the worldwide semiconductor industry is less than $500 billion. The Chinese government is investing more than $20 billion per year. They are doing it in an insightful way. The money is invested as a share of private equity companies who are motivated to invest in semiconductor companies that can make an attractive return on the investment. Many of the startup companies have an optimistic outlook because the Chinese population of over one billion people can drive its own standards for communications and computing.

How well have they done with the investment? Figure 2 shows a comparison of the size of the semiconductor companies that have benefited from the investment.

Average employment of the semiconductor companies in China has grown between 2006 and 2015. The number of companies with more than five hundred employees in 2006 was less than one half of one percent. In 2015, it was 6.1%. The number of Chinese fabless semiconductor companies that had between 100 and 500 employees in 2006 was 9.8%. In 2015, it was 43.3%.

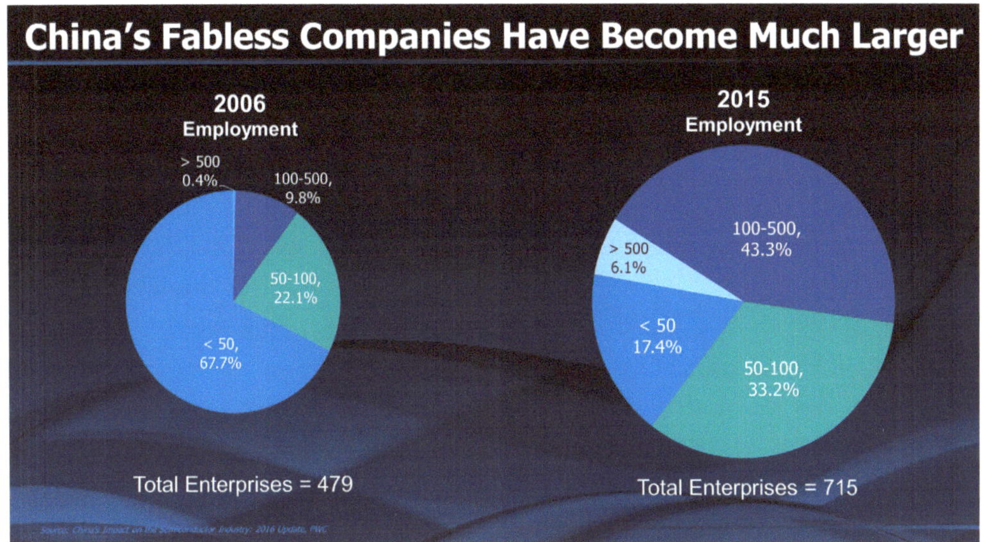

Figure 2. Growth of Chinese fabless semiconductor companies.

This growth provides a challenge for semiconductor competitors in the rest of the world including the U.S. where governments can't afford to provide the kind of subsidies that are available in China. Even so, the problem was manageable with innovative U.S. companies operating in a semi-free market environment for creating new technologies. The U.S., however, created a "Sputnik moment" for China that may have changed the outlook.

Historical Perspective of Export Controls

In 1982, I joined the Technical Advisory Committee of the Department of Commerce that was set up to advise the Department regarding the granting of validated licenses for export of semiconductors and semiconductor manufacturing equipment. Soon I became Chairman of the committee. In the 1970s, U.S. companies like Applied Materials, Lam Research, Novellus, Varian and many more dominated the worldwide semiconductor manufacturing equipment business. Japanese companies like TEL were growing but their base of customers was largely in Japan.

The Export Administration Act of 1979 was one of the contributing factors that changed that competitive situation, although not the only factor since photolithography was increasingly becoming dominated by non-U.S. suppliers. Concern about the military implications of semiconductor capability led to a set of restrictions on semiconductor related exports from the U.S. and its allies to controlled destinations that were enemies of the free world.

Allies like Japan were much more efficient at administration of bureaucratic export control rules. Validated licenses for customers of Japanese manufacturers could reliably be obtained in three days plus or minus a day or two. For the U.S., the time and the variability were much greater. Representatives of SEMI, the principal trade association for semiconductor manufacturing equipment suppliers, also noted that the U.S. Department of Commerce applied different interpretations to the regulations from our COCOM allies.

Victoria Hadfield, Government Relations Manager for SEMI said, "… semiconductor manufacturing equipment exports to Japan were subject to U.S. licensing requirements, despite the fact that there were many sources of competitive products in Japan".[2] She also noted that Japanese fabrication lines had been installed in facilities in China because Japan argued that, despite controls on individual pieces of equipment, there was nothing in the regulations that said you couldn't ship an entire fab line[3].

Companies in places like Taiwan rightly concluded that the U.S. could not be trusted as a supplier because of inflexibility. Delays in licensing also affected spare parts and user manuals since they too required separate validated licenses.

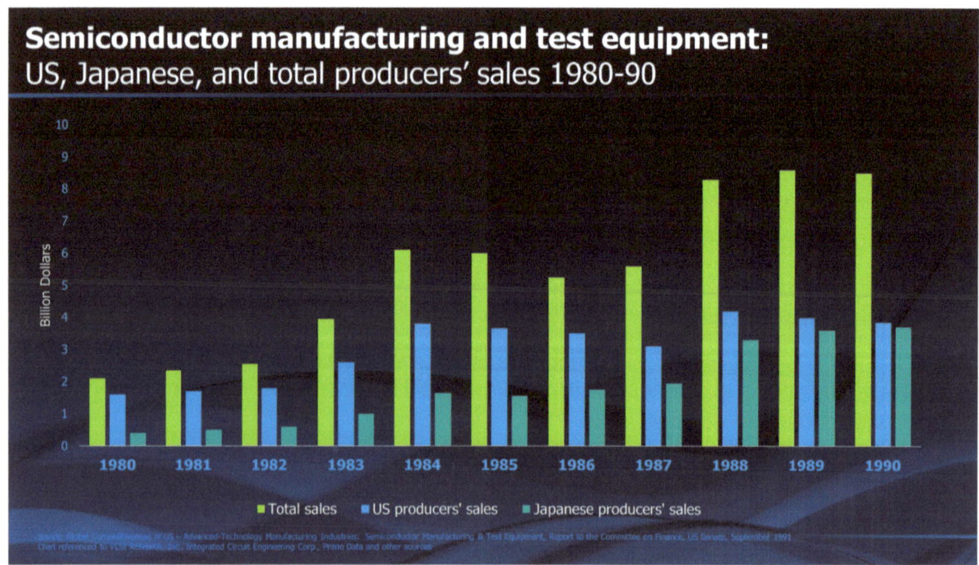

Figure 3. Japan's market share of semiconductor manufacturing equipment increased rapidly after enactment of the Export Administration Act of 1979.

Market share of U.S. companies in the semiconductor manufacturing equipment industry fell from more than 4X Japan's market share in 1980 to about equal shares for Japan and the U.S. in 1990 (Figure 3).

In a sense, the U.S. moved toward a role as the "vendor of last resort" for semiconductor manufacturing equipment sold to controlled destinations and probably even to some destinations that were COCOM allies.

China's Sputnik Moment

ZTE clearly violated internationally accepted export restrictions on Iran. The U.S. reaction was not totally unexpected. What surprised the world was the swift action taken to shut down the free market purchase of semiconductor components in April 2018. The degree of dependence that ZTE had developed upon U.S. semiconductor suppliers made this a life or death issue for the company. President Trump ultimately waived the export restrictions. Although the embargo was lifted on July 13, 2018, the impact of the threat was now apparent to China. Dependence upon U.S. suppliers of semiconductors was no longer a viable strategy.

As the Chinese government continued its move toward dictatorship, rules were imposed upon foreign companies for the opportunity to establish operations in China. Google was prevented from operating in China, as were others. These restrictions became increasingly egregious (such as the threat that U.S. software companies might have to turn over their source code to the Chinese government if they wished to operate in China). As frictions developed in the trade negotiations between the U.S. and China, a U.S. decision was made to embargo exports to Huawei by placing Huawei on the "entity list" in May of 2019. This created an even more untenable situation for China, the worldwide leader in wireless communication technology.

China did the expected. They focused upon developing non-U.S. capabilities for all their components. Since China buys more than 50% of all semiconductor components in the world and uses more than 15% of the world's semiconductor supply in equipment designed by Chinese companies, this is a big problem for the U.S. semiconductor industry. It is probably not reversible. In December 2019, Huawei surprised the world with the introduction of a cell phone containing no U.S. semiconductor components.

Another question is whether other countries will follow suit out of fear that political disagreements with the U.S. could result in an embargo of semiconductor components from U.S. suppliers. As with the Export Administration Act, events in China will lead to a reduction in market share of U.S. based semiconductor suppliers and a loss of their lead in many new technologies.

How long will it take for China to achieve total independence from U.S. semiconductor suppliers? Probably many years, especially for FPGAs and RF components, and it may never be achieved. China's demand is so large that non-U.S. suppliers will probably find a way, given enough time, to displace U.S. suppliers but it's hard for any country to become totally self-sufficient. In addition, China's move toward a closed, controlled society will restrict innovation. That will work to the advantage of the U.S.

While China's direction is not likely to change, we still have the possibility of convincing the rest of the world that the U.S. can be treated as a reliable supplier. Hopefully, there will be policies articulated by the U.S. that convey that confidence and restore the U.S. position as a leader in free trade.

1 https://en.wikipedia.org/wiki/Industrial_Technology_Research_Institute

2 https://www.usitc.gov/publications/332/pub2434.pdf

3 Ibid

Chapter 12: The Future

The content of this book has focused upon predictability of trends in the semiconductor industry based upon past trends, experience and ratios. What about newly emerging applications of semiconductors? After all, the entire history of the semiconductor industry is driven by emergence of new applications.

Artificial Intelligence

One of the most exciting new applications affecting semiconductor technology is the broad adoption of AI-related analytics. AI is not a new technology. Figure 1 is the cover of High Technology magazine in July 1986. I am the person on the left and George Heilmeier, former head of DARPA, is the one on the right. We tried hard in the 1980s but the infrastructure had not developed to a level where AI would provide profitable opportunities

Figure 1. Artificial Intelligence technology heavyweights of the 1980s.

What's different today? In the 1980s, we lacked the computing power to handle big data. We didn't have very much big data to analyze partly because there was no internet of things. More sophisticated algorithms were needed. Most of all, there were no obvious near term ways to make money using AI.

Today we have overcome all these limitations. AI and machine learning have taken on a life of their own. They have become limited, however, by the processing power available. Traditional von Neuman general purpose computing architectures are inadequate to handle the complex AI algorithms. The result: a new generation of computer architectures is evolving.

Figure 2 shows the trend in venture capital funded fabless semiconductor companies in recent years. In 2018, a new record of $3.4 billion total investment was set, far above the previous record of $2.5 billion in the year 2000.

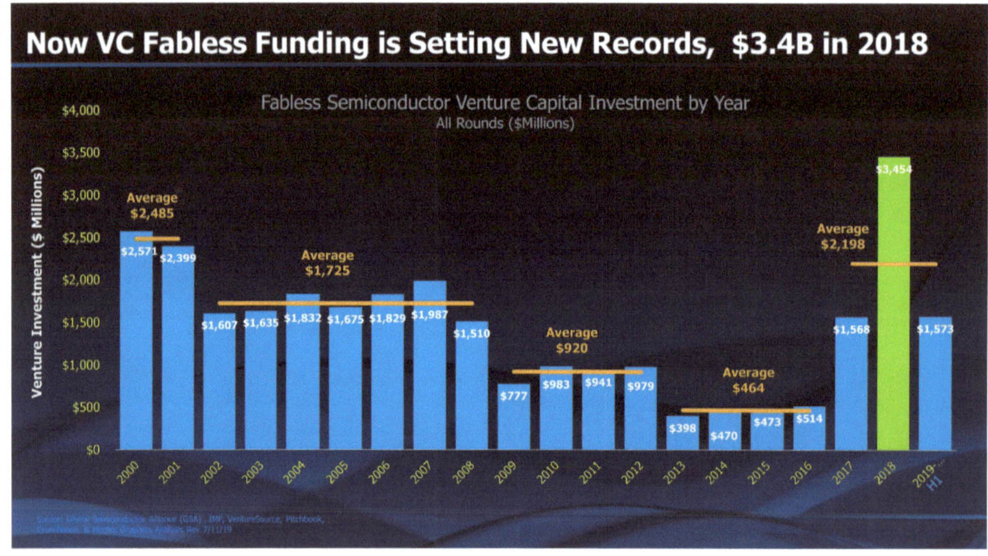

Figure 2. *Venture capital funded fabless semiconductor startups.*

Venture capitalists have been focused on social media and software companies over the last twenty years. Where is all this new money going? The answer can be seen in Figure 3. AI and machine learning have dominated the first three rounds of fabless semiconductor startup investment by venture capitalists since 2012 with $1.9 billion invested.

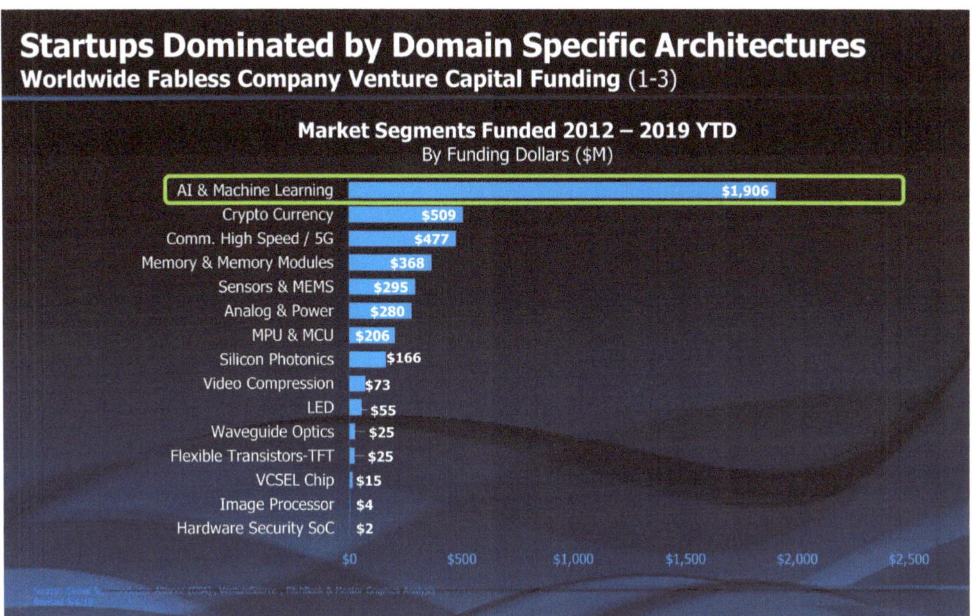

Figure 3. AI and machine learning have dominated the first three rounds of fabless semiconductor startup investment by venture capitalists since 2012 with $1.9 billion invested.

What kind of chips are being funded? The largest share is in the area of pattern recognition. Chinese investments in facial recognition chips developed at companies like Sensetime and Face++ constitute a very large share. There are seventy-five other companies developing chips for pattern recognition that have been funded by venture capital. These include companies focused on pattern recognition for audio, smell, medical diagnostics and other applications.

Beyond pattern recognition, the largest share of new fabless semiconductor companies are being funded for data center analytics or edge computing.

Edge Computing

Intelligence historically flows downhill (Figure 4). In the 1960s, mainframe computers dominated our computing capacity. Dumb terminals connected us to our mainframe computing power. By the 1980s, minicomputers were well established as an intermediate computing layer between the user and the mainframe. Twenty years later, the personal computer became the local computing resource.

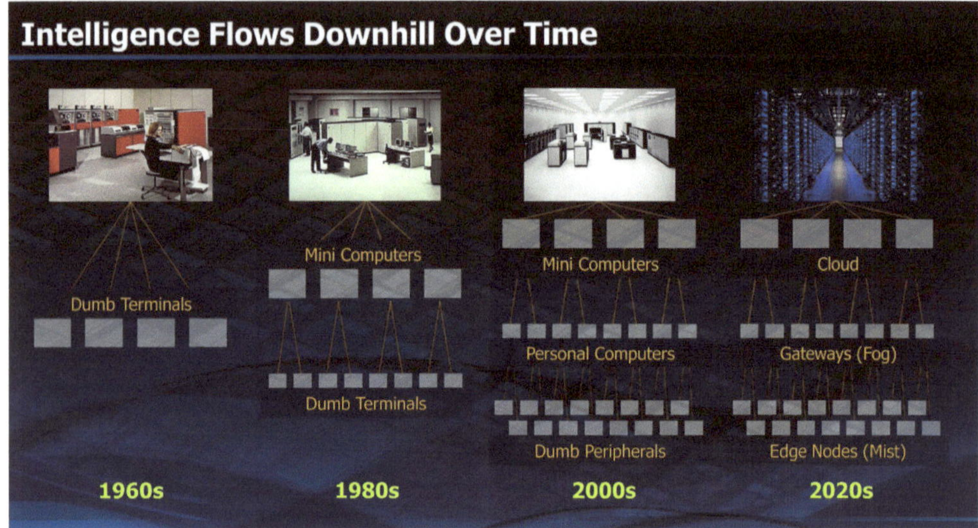

Figure 4. Intelligence flows downhill.

In another twenty years, the current environment has evolved. Large cloud-based server resources handle the heavy computing but in between us and the cloud is the fog made up of gateways that collect, aggregate and locally process data. Beneath that layer are the edge nodes in the mist, collecting and pre-processing the data.

As time passes, the lower layers will inevitably gain more intelligence as semiconductor technology allows us to build more intelligence into the local nodes. Those nodes will become increasingly complex as they incorporate disparate technologies – analog, digital, RF, MEMs, etc. (Figure 5). This creates major design and verification challenges. Most of the largest revenue in EDA history came from digital logic and memory. Edge nodes will require mixed technologies. Simulating digital logic connected to analog, RF and other technologies is not easy. A whole new family of EDA tools is required.

5G Wireless Technology

In the next decade, wireless communication will move to the next generation of technology, 5G. This transition is more significant than past generations. It affects a wide variety of the infrastructure beyond hand-held wireless communications. Significant impact will be felt in applications involving industrial control, non-real time automotive analytics, urban infrastructure and much more.

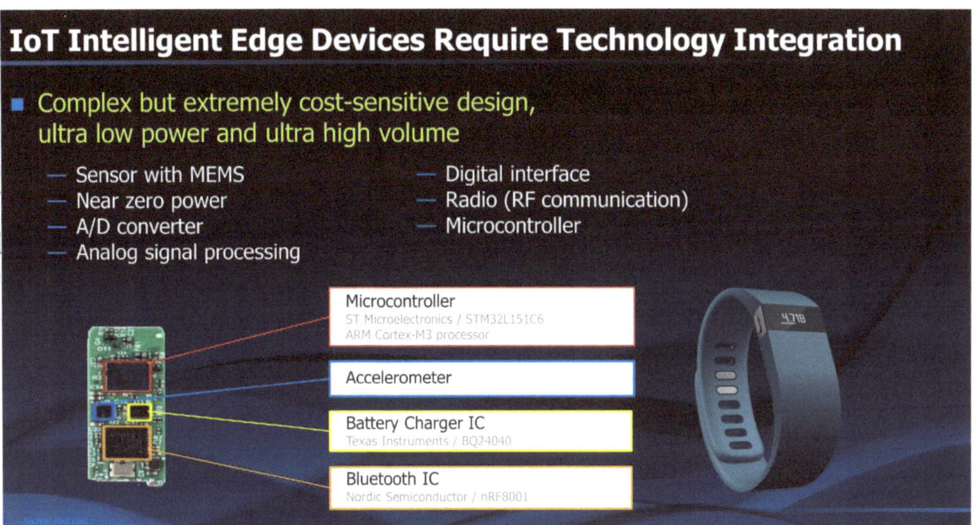

Figure 5. Diverse technologies like digital, analog, RF and MEMs will be required as edge nodes become more intelligent.

One of the great opportunities for the semiconductor industry is the increased number of base stations required to support the infrastructure of 5G and the larger number of antennas in a phone. The number of semiconductor components required will grow dramatically as the world builds a 5G infrastructure. Connectivity to the cloud makes a wide variety of capabilities possible, especially in the factories of the world. Gateways, which already generate more than three percent of worldwide semiconductor revenue, will be needed.

This connected world will be dependent upon more semiconductors for communications and computing. For many years the semiconductor industry measured its revenue from the computing and communications industries which were each about 35% of the total. Now the two are indistinguishable. Seventy percent of the revenue in the semiconductor industry comes from one or the other or a unique combination of both.

Automotive Applications

During the last ten years, sales of semiconductors for automotive applications has increased to about 12% of the total semiconductor market as the electronic content of vehicles increased. Some traditional electronic functions like engine control will not be needed in electric vehicles but there will be new requirements as well as the continued growth of infotainment, communications and automotive driver assistance (ADAS) that require electronic controls.

The number of companies planning to build electric cars or light trucks has now grown to 463, more than half of which are based in China (Figure 6). Two hundred eleven companies have announced autonomous driving programs.

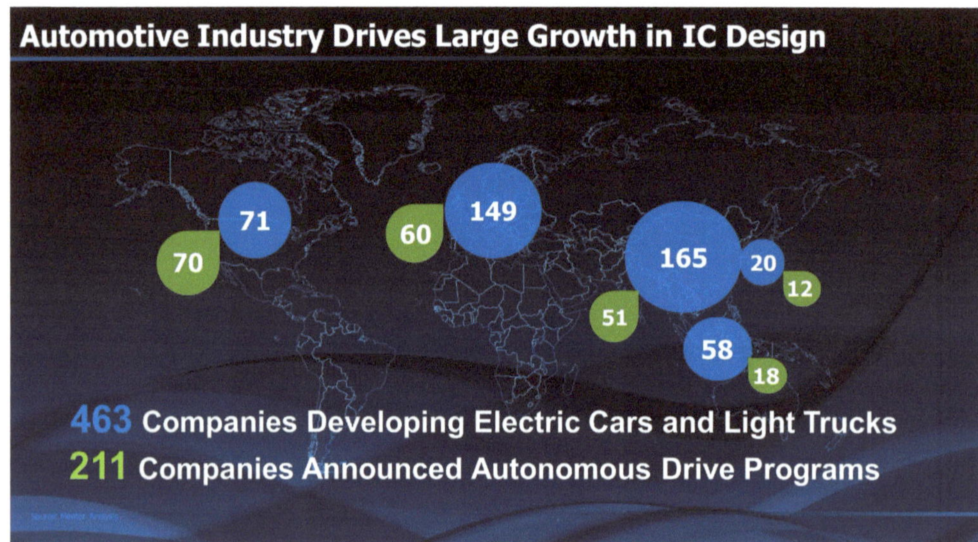

Figure 6. As of June 2019, 463 companies have announced intent to introduce electric cars or light trucks. 211 companies have announced autonomous drive programs.

This number of suppliers is not needed and many, or even most, will drop out as we move closer to production. Meanwhile, one would expect an incredible bubble in demand for automotive ICs followed by a temporary decline.

It's likely that no more than a dozen companies will lead the way in supplying the complex image processing subsystems required for autonomous vehicles. It's difficult to predict which ones will succeed but likely that companies that have not been traditional automotive OEMs will make up most of the total.

Other Predictable Futures

Lots of other technologies offer promise for growth. Quantum computing is interesting because it has some capabilities like encryption that are not solved easily through other means. The time lag for technologies like this tend to be longer than the evolutionary ones but they will eventually emerge in some form. Highly secure forms of encryption like fully homomorphic encryption still require six orders of magnitude more performance than current chips provide but there are machine learning techniques that may achieve this capability within the next decade or much sooner. For many, designing secure server centers and networks starts with the assumption that the network has been breached and that the innovative structure of data will provide the next line of defense.

Blockchain technology also offers a wide variety of opportunities for semiconductor design innovation. Beyond Bitcoin mining, which caused a temporary boom in custom IC design, there are a wide range of applications that will benefit from incorporation of blockchain technology, including much of our world of contract agreements and documentation. There's also a reasonable probability that China will establish its own blockchain based cryptocurrency as a challenge to the domination of the U.S. dollar as the reserve currency of the world.

The history of the semiconductor industry is driven by major new applications. Waves of growth were initiated by defense electronics in the 1950s, mainframe computers in the 60s, minicomputers in the 70s, personal computers in the 80s, laptops in the 90s and wireless communications in the most recent twenty years. Each wave has been accompanied by the emergence of new semiconductor competitors followed by a shakeout that leaves one supplier dominant and shuffles the top ten ranking of companies by revenue (Figure 10 in Chapter 5).

At the same time, the semiconductor industry, like most industries, moves back and forth from standardized versus customized solutions. This has been referred to as "Makimoto's Wave" after Tsugio Makimoto, former CEO of Hitachi Semiconductor, who popularized the phenomenon. As we move into the third decade of the twenty-first century, the semiconductor industry is moving into a customization wave.

Standard von Neuman computer architectures that operate on a string of standard instructions have dominated the computer and semiconductor industries. Architectures like the Intel 808X and ARM RISC will continue. Domain-specific architectures tailored for specific tasks like facial recognition are emerging. There will be dozens more as AI and machine learning usher in new opportunities.

What should we consider as the future possibilities for the semiconductor industry? As we showed in Chapter 4, the semiconductor industry is likely to grow through evolutionary means through about 2040 or when demands for lower power or higher performance usher in a new technology for information "switching". Carbon nanotubes, bio-switches, or many other possibilities could fill in the switching learning curve of Figure 5, Chapter 3. Chances are that this "switch" will happen gradually as the need arises for a new application. In addition, non-silicon materials like Gallium Nitride, Silicon Carbide and other materials will take on increasingly important roles driven by need for characteristics like larger band gaps, i.e. roles like power switching, microwave communications and existing ones like solid state lighting.

Just as steel is still a primary material for construction one hundred fifty years after the booming growth of the steel industry, semiconductors will be at the foundation of business and technology growth for a long time. Those of us who participated in the last fifty years of exciting growth of semiconductors are still surprised when we see our "mature" industry generate another wave of growth to accompany an emerging application. I'm confident that there will be many more to come.

The Innovators

Throughout most of the semiconductor history, progress has been limited primarily by the number of trained, innovative people available to create new technologies and applications. Although much of the student interest in universities has moved from hardware to software in recent years, there seems to be a rejuvenation of hardware interest as artificial intelligence applications require large increases in performance at reduced power. Where will we find the talented people to keep up the stream of innovations?

One source of talent and funding resources is coming from a fundamental shift in semiconductor R&D and production. Figure 7 shows the shift in purchases of wafers from foundries.

Ten years ago, fabless semiconductor companies bought nearly 80% of all wafers, with integrated device manufacturers like Intel, TI, Samsung, etc., purchasing most of the rest. During the last five years, the percentage of wafers purchased by systems companies has grown at a 70% compound average growth rate. Apple, Google, Facebook, Amazon and many more like them, as well as the entire automotive industry, have established leading edge

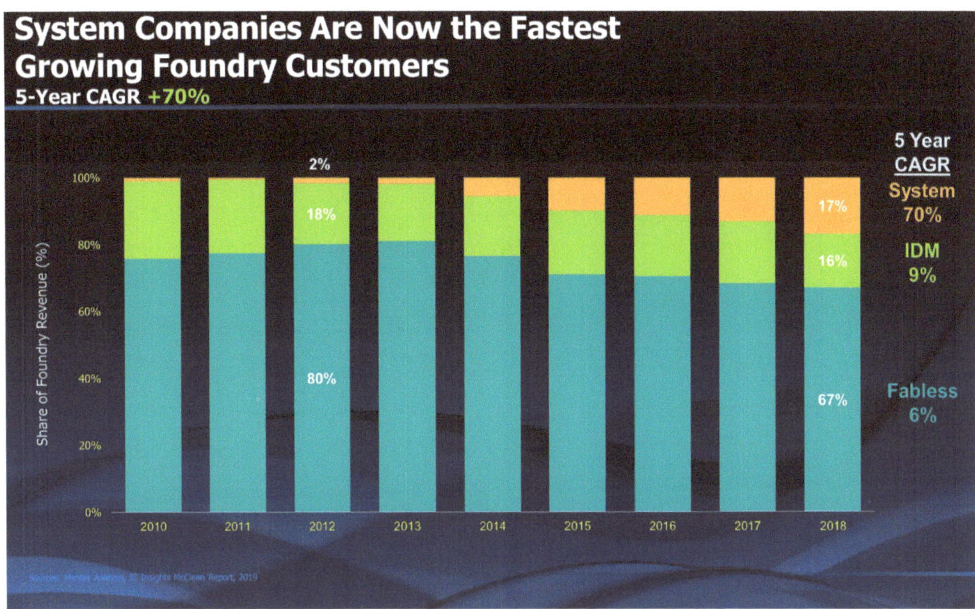

Figure 7. Increasing share of foundry wafers purchased by system companies.

design groups and purchasing capability for their chips. For the information technology companies, designs include custom servers, compute accelerators, IoT networks and new types of data analytics. This trend has caused a major increase in EDA industry growth rates as well as increasing the people dedicated to semiconductor innovation.

At the same time that systems companies add to the base of chip design capability, the massive investment by China, in its effort to become more self-sufficient, will bring tens of thousands, or possibly millions, of new innovators to design and manufacturing innovation.

This brings us to the question of elasticity of semiconductor resources. With software, the growth in opportunities led to accelerated involvement of millions of engineers and scientists. Could the same be true for semiconductor designers? Probably not. The cost of prototyping, system engineering, embedded software development and wafer fabs limits the number of designs that the world is likely to undertake. Availability of design ecosystems in the cloud, however, could bring design capability to many who could not previously afford to put their innovations into custom silicon. Investment funds are being established to offer essentially free access to design tools and silicon prototypes during the development phase of promising startups.

A decade ago, many suggested that integrated circuit design would become a discipline limited to the large companies with the financial resources to spend tens, or hundreds, of millions of dollars on a single design. Now, thanks to efficiencies in design automation, we see dozens of new startups designing chips at complexity levels that challenge the maximum size of a reticule. They are also putting together chiplets into packages that keep the transistor count growing and maintaining the same traditional steep learning curve for cost and power dissipation.

No country has a monopoly on talented, innovative people. The U.S. has been a mecca in the past for innovators from around the world to use their talents, achieve their dreams and, in some cases, become very wealthy. It's my hope that the U.S. can continue this role in the future.

www.ingramcontent.com/pod-product-compliance
Lightning Source LLC
Chambersburg PA
CBHW040921110726
48006CB00001B/18